Books by Margaret Mead

COMING OF AGE IN SAMOA
GROWING UP IN NEW GUINEA
THE CHANGING CULTURE OF AN INDIAN TRIBE
SEX AND TEMPERAMENT IN THREE PRIMITIVE SOCIETIES
AND KEEP YOUR POWDER DRY
BALINESE CHARACTER: A PHOTOGRAPHIC ANALYSIS *with Gregory Bateson*
MALE AND FEMALE
GROWTH AND CULTURE: A PHOTOGRAPHIC STUDY OF BALINESE CHILDHOOD *with Frances C. Macgregor*
THE SCHOOL IN AMERICAN CULTURE
SOVIET ATTITUDES TOWARD AUTHORITY
NEW LIVES FOR OLD: CULTURAL TRANSFORMATION—MANUS, 1928–1953
AN ANTHROPOLOGIST AT WORK: WRITINGS OF RUTH BENEDICT
PEOPLE AND PLACES
CONTINUITIES IN CULTURAL EVOLUTION
ANTHROPOLOGY: A HUMAN SCIENCE
ANTHROPOLOGISTS AND WHAT THEY DO
FAMILY *with Ken Heyman*
THE WAGON AND THE STAR: A STUDY OF AMERICAN COMMUNITY INITIATIVE *with Muriel Brown*
THE SMALL CONFERENCE: AN INNOVATION IN COMMUNICATION *with Paul Byers*
CULTURE AND COMMITMENT
RAP ON RACE *with James Baldwin*
BLACKBERRY WINTER: MY EARLIER YEARS
TWENTIETH CENTURY FAITH: HOPE AND SURVIVAL
RUTH BENEDICT: A BIOGRAPHY

Books Edited by Margaret Mead

COOPERATION AND COMPETITION AMONG PRIMITIVE PEOPLES
CULTURAL PATTERNS AND TECHNOLOGICAL CHANGE
PRIMITIVE HERITAGE: AN ANTHROPOLOGICAL ANTHOLOGY *with Nicholas Calas*
THE STUDY OF CULTURE AT A DISTANCE *with Rhoda Métraux*
CHILDHOOD IN CONTEMPORARY CULTURES *with Martha Wolfenstein*
THE GOLDEN AGE OF AMERICAN ANTHROPOLOGY *with Ruth Bunzel*
AMERICAN WOMEN *with Frances B. Kaplan*
SCIENCE AND THE CONCEPT OF RACE *with Theodosius Dobzhansky, Ethel Tobach, and Robert E. Light*

Books by Ken Heyman

WILLIE *with Michael Mason*
POP ART *with John Rublowsky*
FAMILY *with Margaret Mead*
THIS AMERICA *with Lyndon B. Johnson*
COLOR OF MAN *with Robert Cohen*
THE PRIVATE WORLD OF LEONARD BERNSTEIN *with John Gruen*
THEY BECAME WHAT THEY BEHELD *with Edmund Carpenter*

WORLD ENOUGH

WORLD ENOUGH

Rethinking the Future

Words by
Margaret Mead

Photographs by
Ken Heyman

LITTLE, BROWN AND COMPANY
BOSTON — TORONTO

FIRST EDITION

T 03/76

Grateful acknowledgment is made to Léonie Adams for permission to reprint lines from "A Gull Goes Up" from *Poems: A Selection* by Léonie Adams. Copyright 1925, 1953, 1954 by Léonie Adams. (This poem was originally published in *Those Not Elect*, Robert M. McBride & Company, New York.)

LIBRARY OF CONGRESS CATALOGING IN PUBLICATION DATA

Mead, Margaret, 1901–
  World enough.

  Bibliography: p.
  1. Ethnology. 2. Geography. I. Heyman, Ken. II. Title.
GN378.M42   301.2   75-25680
ISBN 0-316-56470-2

Design by Barbara Bell Pitnof

Published simultaneously in Canada by Little, Brown & Company (Canada) Limited

PRINTED IN THE UNITED STATES OF AMERICA

This book is dedicated to
the future of my Granddaughter,
Sevanne Margaret Kassarjian
and of my Great-Goddaughter,
Katherine Suzanne Metraux

MARGARET MEAD

the future of my Children,
Jennifer Heyman
Timothy Heyman
Christopher Heyman
Jason Heyman
and Amanda Heyman

KEN HEYMAN

Had we but world enough, and time . . .

ANDREW MARVELL
(1621–1678)

Acknowledgments

The authors acknowledge with appreciation the skillful work of their editor, Chauncey G. Olinger, Jr. They also are grateful for the help of Natalie Smith, assistant to Mr. Heyman, and to Marie Fatte and Erica Leone of Meridian Photographic Laboratories, in the preparation of the photographs for this book. Both authors express their gratitude to Wendy Heyman for her generous patience and understanding.

————

Some of the material included in the text of this book was developed on the following research projects supported by grants from the National Institute of Mental Health: M-2118, "Reconnaissance Research on Balinese Mental Health," 1957–59, and MH-3303-01, "The Factor of Mental Health in Allopsychic Orientation," 1961–65.

————

The quotation on page xxii is from *People and Places* by Margaret Mead (Cleveland, Ohio: World Publishing Co., 1959), p. 309; on page xxiv from "Stanzas from the Grande Chartreuse," by Matthew Arnold; on page xxiv from *Henry IV: Part I*, Act 1, Scene II; on page xxviii from Heraclitus, Fragment 91, cited in *Ancilla to the Pre-Socratic Philosophers* by Kathleen Freeman (Cambridge, Massachusetts: Harvard University Press, 1957), p. 31; on page 43 from "The Chambered Nautilus" by Oliver Wendell Holmes; on page 161 from "The Garden of Proserpine" by Algernon Charles Swinburne; and on page 202 from "A Gull Goes Up," *Poems: A Selection* by Léonie Adams (New York: Funk & Wagnalls Company, 1954), p. 106.

Contents

I
The Dream of Technological Salvation

II
The Waiting World

III
The Failure of the Dream

IV
Beginning Again

List of
Photographs

I

The Dream of Technological Salvation

II

The Waiting World

III

The Failure of the Dream

IV

Beginning Again

Introduction

When my daughter was two years old she asked for the world, and I gave her a globe — a small one that she could turn with her hands. When she was four her father went to Ceylon. She followed the course of his troopship on the globe, and, when a cable came announcing that he had arrived, said, "That means they wouldn't sink the ship Daddy was on — while he was on it."

Ken Heyman and I met in 1954 when he entered my anthropology class at Columbia University. Two years later, while he was on an around-the-world photographic journey, we planned for our paths to cross in Bali. There we were joined by I Made Kaler and Ida Bagoes Made Djelantik, two Balinese friends. When Gregory Bateson and I worked in Bali from 1936 to 1939, I Made Kaler was our Balinese secretary; Ida Bagoes Made Djelantik, the son of a raja, was a youth then, but now he was in charge of the health services for all of the people of Bali. They went with us back to the village of Bayung Gedé, where Ken photographed many of the same people — those who had been children were now grown to adulthood — whom Gregory Bateson had photographed twenty years before. In a village where there had been only a handful of people who could decipher the traditional Balinese script, the schoolchildren now drew accurate diagrams of *Sputnik*, which had passed, so high, over their heads.

Then, in 1957, Velma Varner, a brilliant young editor of children's books for the World Publishing Company, asked me to do a book about the peoples of the world. The book, *People and Places*, was illustrated by artists' drawings of our long human past, of the Eskimos of the far North, of ancient palaces in Africa, and of Indians, charging their enemies, mounted on horses that Europeans had brought to America, and by Ken's photographs of the Balinese, set beside Gregory Bateson's earlier pictures. The children I had studied in the late 1930s were grown now; they were going to school, becoming part of the modern world.

The last picture in that book showed university students preparing to launch their own experimental two-stage rocket. The closing sentences were:

Once more we are entering a period in which men will have to give their whole attention to what they are doing and in which the safety of the whole group will depend on men and women who, as boys and girls, learned that life in the twentieth century is like a parachute jump: You have to get it right the first time.

When the American Anthropological Association met in Mexico City in 1959, Ken and I went to visit some colleagues of mine working in a nearby village. From there they took us to another village, where Ken was allowed to photograph young bulls trying out for the ring and the preparations for Christmas festivities, with people washing pots and pans in the stream that ran through the village.

Then, in 1961, Ken said to me, "I want to do a book about the family. I think 'family' expresses what I want to say about people." (He had three children then, and my daughter had recently married.) He went out to parts of the world where he had been before and to parts where both he and I had been, taking pictures of mothers and fathers, old grandmothers and rollicking small children — and children who did not have enough to eat or a place to sleep. When he returned, we spread the pictures out for the English anthropologist Geoffrey Gorer to see and he said, "But where is the family?" And true, there were fathers and mothers and brothers and sisters, but hardly a single family. So Ken traveled again to find families *together* — a harder task, perhaps, because a family together is more self-conscious about the ties that bind them than is a mother or a father alone with a child. But he did find and photograph many family occasions and *Family* was published in 1963. The thing I enjoyed most about that book was that there was something in it to make a year-old baby smile and something to give tranquillity to the weariest old eyes, while those in between could find echoes of the warmest moments in their lives.

While these years passed, Ken was growing in stature as a world-renowned photographer. He had become a member of Magnum, had worked for *Life* and *Look* and many other magazines, and had exhibited his photographs on three continents. In 1964, we decided that we might do another book, this time about particular families who were living on the edge of change. This we thought would bring together Ken's interest in the beauty of peoples who still lived close to the earth and worked with their hands, and my experience with Oceanic peoples whose lives were changing so rapidly; and it would bring home (especially to Americans) a sense of the widespread change that was happening here to small farmers, black and white, north and south, east and west, as enormous farm machinery replaced the reaping hands of the past.

We had already photographed a Mexican family, and now we planned for Ken to join me in Sicily, during a break in an assignment to photograph Elizabeth Taylor while *Cleopatra* was in production. We were helped by a student of mine who had returned to the little mountain village from which both her father and mother had come. Through a terrible winter of earthquake and epidemic, she had struggled to make friends with her relatives and find a way of studying the intellectual development of children still caught in the toils of an ageless peasant way of life. She had found a school where an intrepid priest had gathered together some of the brightest boys in the countryside to help them make the leap into the modern world. Ken met me there, and we found a family who still threshed their bean harvest with a donkey walking in a circle and baked their bread in a shared outdoor oven, but who, nevertheless, had a television set in their living room.

After this, Ken went back to Bali, to the mountain village of Bayung Gedé — where he and I had worked in 1956 — to photograph a Balinese family. By then, however, Bali had undergone terrible changes: first, an abortive civil war had ended in a massacre — the top children in the school in Bayung Gedé, who had been learning about *Sputnik*, were all killed along with their teachers, who had belonged to the losing party; then, a terrible volcanic eruption had destroyed worshiping thousands who had come to celebrate the hundred-year festival of the gods who dwelt at the top of the mountain, and a third of the arable land had been ruined. Ken found the man who had worked for Gregory and me in 1936, and who had gladly stepped forward to set up cameras for us in 1956. In 1966, he was now the father of eleven children, plowing in the same way, with the same oxen, his life somehow unchanged by massacre, the volcanic eruption a few miles away, or by the jet planes that flew overhead.

An anthropologist friend found us a young farm family in Vermont, whom we photographed; this young couple had moved into an old parental house, with a new television set and shiny plastic chairs and table — the husband's parents had taken the old-fashioned furniture to a brand-new prefabricated bungalow. The young mother kept the television set on most of the day, but with the sound off — a new kind of silent window on the world. And Ken found a black farmer in Louisiana who was making a go of it in a rapidly changing situation in which small farmers were being displaced by agribusinesses where the land permitted the use of the big new machines, but in which plots that were too small and hilly were allowed to go back to wilderness.

We now had five families — one each in Bali, Sicily, and Mexico, and two in the United States — families with roots in the great distinctive areas of the world, Asia, Africa, Europe, the Americas. It had seemed like a good idea; and yet, somehow, it seemed empty. Although I had helped make the Canadian Film Board documentary *Four Families* — Japanese, Indian, French, and Canadian — only ten years before, those four families — so different in all the contexts of their lives, but so human in the scenes of mothers putting their babies to sleep — had, together, seemed to give a sense of our common humanity. But when Ken and I came to do the five families book, somehow the detail didn't carry the meaning we were looking for; there was something missing in the linkage of the past with the changing present; it all seemed lifeless, as if ". . . between two worlds, one dead, / The other powerless to be born."

We laid the pictures aside; regretfully, because it seemed a poor return for the kindness with which our photographic visits had been received. This was 1971. I know now that this was the year that saw the death of the last flicker of hope that technological change, planned so optimistically after World War II, would bring peace and prosperity to all the peoples of the world. We had, indeed, come to the end of an era, the quarter of a century after World War II, which we had entered with such high expectations. The five families, perched precariously on the precipitous edge of change, belonged to a past that now had to be understood as something very different from a prelude to the future we had thought would follow. In 1940, Marjorie Nicholson had used as a text in a commencement address at Wilson College: "We must '. . . pay the debt . . . [we] never promised.' " Now we see that the attempt to make our debt to the world into a promise has failed.

In 1965, Ken was commissioned to do the photographs for *This America*, a vision of the Great Society which Lyndon Johnson had hoped would be his special contribution, in the great tradition of Franklin D. Roosevelt, to America. Ken saw and photographed this country in the spirit of the New Deal: poverty, rural toil and neglect, desolate city streets, joined to the expressed hope that all could be remedied — if we just had the will to do so. By the time the book was published, the dream had soured, Johnson was a disgruntled and unhappy man, and the campuses of the country were seething with rebellion against the U.S. involvement in the Vietnam war. Johnson's bursts of rhetoric echoed hollowly against the background of that unhappy war.

I find it significant that here Ken draws very little on those scenes of destitution and promise from his journeys up and down the country. Pictures of a destitution that can be remedied by a Poverty Program and an Office of Equal Opportunity, while American planes were bombing fields that had been tilled so many thousands of years before the first plow broke the American plains, somehow no longer convey the same message. The great, fertile, safe fields of the United States present a strange contrast to the ruined landscape of Vietnam, pitted by craters, given over to the creeping destructiveness of the bamboo that springs up when the protective forests are destroyed. And even the little module for landing on the moon he photographed stark and somber against the sky.

Then, in the winter of 1974, Ken came into my office and said, "I want to do a book about the world." He said it as simply as he had said more than a decade before, "I want to do a book about the family." Through my mind went one of those haunting contrapuntal echoes that infest our minds today, "Tomorrow the World." It was a dictator who said, "Today Germany, Tomorrow the World." Ken was saying, "First the Family, Tomorrow the World." Quite a different theme, of course. And I said: "When do you want to start?" Later, he said that he felt it was odd that I should have agreed so readily, while I felt it was not surprising, that through all our unsuccessful efforts to focus on families, this intention had been there: Tomorrow the World.

An impossible assignment? Perhaps. We spoke of his returning to places where we had worked together and of his going to other places: Japan, Hong Kong, Nigeria, and Venezuela, where he had gone for the Alliance for Progress, when it too carried as gilded a rhetorical vision as Lyndon Johnson's Great Society. We wondered whether we could find the family, in that little Mexican village, whom we had met on that 1959 Christmas, and who would be alive in the remote Balinese village of Bayung Gedé. We would make the book a study of the changing world, setting side by side pictures of the same people grown a decade older.

I had tried before to make the passage of time in individual human lives a kind of living lesson that could be easily absorbed by children newly come to the idea of a calendar and chronology. In the New Guinea village of Pere in 1953, I had made three calendars for the schoolchildren, one showing the estimated age of the earth and the appearance of humankind, and a second with the dates they had recently learned: the birth of Christ, the discovery of America, the discovery of Australia, the discovery of New Guinea, World War I, World War II, and 1946, when their own new social order had begun. For the third calendar, I made a scale in decades, identifying each decade by the birth of a member of the village — with the oldest man going back to the 1890s — so they could fit biological time into the long chronologies of the age of the earth and the chapters of recorded history with which they would soon have to deal. This had worked well and the European calendar had come to life for them. Perhaps, if we could show the same people over a span of years — children grown to adulthood, hands gnarled with heavy work, women's once uplifted breasts sagging from having nursed many children, tractors succeeding oxen in the fields, pumps for the wells where women once drew water — we would be able to bring to life, for the inhabitants of the last quarter of the twentieth century, the changes that were going on in the world.

So Ken went out again, to the peoples of five continents of the world. He found them and brought photographs back and spread them on his living room floor. Then his oldest child asked him what this book was to be about, and he started to explain that these were pictures of change. Suddenly he realized that many were not pictures of change at all. What he had recorded was unchangingness, lives still untouched by the changes believed to be going on, hands as scarred as their parents' and grandparents' had been, and children whose ribs showed through their taut skins even more sharply than before.

When we looked at the pictures together, the theme of this book became clear: the extravagant hopes that high technology would solve all the problems of poverty, hunger, disease, and ignorance, the belief that the mechanized agriculture that had worked so well in Kansas and Nebraska, in England and Australia, could be used equally well on the thin soil of the tropics, the belief that the style of political organization in the industrialized countries could be grafted onto the traditional behavior of the rural and tribal peoples of the world, the belief that somehow the rest of the world should stand still, while the economically disadvantaged caught up — it has all turned out so differently. It was clear that we needed some different hopes.

Our overburdened memories are strangely blurred today. The very quantity of the pictures from the press, film, and television that crowd into our minds means that the image of each new event is continually jostling for room beside the images of events that happened only yesterday. In a New Guinea village, events are remembered for generations; a house burns down and no new house is built, an ongoing reminder of the fire. When two people pass the empty site, one makes the comment, "That was the fire that a small boy, the son of Purudimi, started in anger," or, "That was the place of the home of the one-legged man." When I return to find people I first met over thirty years ago, they remind me, "It was my father . . ." or, "It was my uncle who carried you, trussed up like a pig, over the mountains to Alitoa." If on a return visit, I stumble unrecognizingly over a name, someone is sure to say, "We didn't call him by that name when you were here; you called him by another name, the one his uncle from a far island gave him." It is like this in the villages, where the memories of drought and flood, repeated and relived, parallel the deep lines in the people's weather-beaten faces.

But in the modern world, headlines scream even if no one in the day's news is crying for help; headline writers search for murder and arson and rape, if not in this city then in a small town in another state, if not in this country, then far away on the other side of the world. Events of extraordinary difference in scale — a tiger in the luggage compartment of a 747 and the crash of a plane with the loss of its passengers, the outbreak of a civil war and the toppling of a revolutionary regime, one bullet gone astray in New York and a load of bombs dropped by mistake on allies, a new kind of tooth powder guarantees to make the breath of the old sweet for the withdrawing young and a child-proof bottle top that neither old nor young can open is invented — all these disparate, incomparable items seem almost to cancel one another out. The kind of festivals that once sustained the poor and unfortunate and kept their lives bright with cherished memories are gone. Young people fill their rooms with giant posters and with pictures the size of postage stamps; but these differences in size don't represent a parallel scaling of values: the little picture may be of a hero, the big picture of so passing a fancy that — away from the room — the owner may be unable to remember which quartet it is that stares down from his ceiling.

Young artists and poets have busied themselves with playing with this fragmented world, putting the bits together in new ways, cut out, stuck together with glue, blown up, cut down, half disappearing. The television images move faster and faster, and our eyes learn to see what we would have missed altogether a decade before. Surely, from these experiences new kinds of memories and new ways of handling abstract ideas, earlier in life, with more of the body involved in thinking, will develop.

But for the people who are here now, even for those who have grown up on television in this kaleidoscopic world, it is hard to remember. Thus, it was less than ten years after World War II that a poll showed that over half the secondary schoolchildren in Switzerland did not know who Hitler was; for a while, after *Sputnik* went up, almost everyone could spell "satellite," but this exactitude passed quickly.

In the 1940s, in the United States, a candidate for the presidency, Wendell L. Willkie, wrote a book called *One World*. He traveled far and wide and wrote glowingly, if without distinction, of the freedom that was to come to everyone, everywhere. John Grierson, the father of the documentary film, took scraps of newsreels and strung them together to make a commentary, exalting the spread of freedom and democracy, that contrasted vividly with the contrived filming of the Nuremberg mass rally saluting Hitler. Grierson's film ended with Chinese refugees crawling on their hands and knees to safety. I used to show it to my classes in the early 1950s; after a while, I stopped showing it, because the hopes that it expressed were no longer meaningful. Now if I showed it, no one would know which Chinese they were or toward which promise they were crawling, as the years of Chinese struggle, of revolution and counterrevolution, melt together in our overheated minds.

There is hardly a part of the world, a nation, or a city which has not come in and out of the news, become temporarily associated with war or earthquake, assassination or rendezvous, shone brightly or menacingly on some explanatory chart or map, and then disappeared, leaving us with only a slight resonance in our minds. Each day, almost each hour, as a car radio is turned on, or as a child pulls the knob on a television set, some new place in the world, or some new human face, rises from obscurity and anonymity into salience. As this happens, we ourselves change our position, align or realign ourselves as concerned or indifferent, hostile or friendly, part of or alienated from the new event. We know, have heard, have seen, a little bit about almost everywhere — but which, or where, is dim. It gives us a sense of false familiarity, like the faces of strangers with whom one has just shared a long wait in a foreign airport. About 500 B.C., Heraclitus wrote what is becoming for us a truism: "It is not possible to step twice into the same river."

A pressing task for the present generations is to learn how to handle this maelstrom of swirling images and words so that they can think about it in some coherent and continuing fashion. Our own identities depend upon where we place ourselves in time and space and how we perceive ourselves with respect to the known and the unknown, the familiar and the strange. Men and women once lived all their lives in small villages where every face and every name was known, even the names of the dead, which one was forbidden to pronounce. Now, today, the urgent need is to find ways of knowing this vast panorama which our new means of transmitting images and sounds as well as of swift travel have opened up to us.

This book is a new experiment, an attempt to create a "macroscope," a way of looking at and understanding something that is large, relatively unknown, and relatively unknowable, which, nevertheless, we need somehow to know. We are just beginning to think about the need for such a way of seeing to match the wonders of the microscope, which approaches the almost infinitely small, and the telescope, which brings us closer to the almost infinitely large. And the need is clear; at the present, we are more and more faced with the unmanageably large: billions of cells in the brain, billions of human beings on the earth, millions of tons of wheat, planes that travel faster than sound, billions of dollars suddenly accumulating in the coffers of small sheikdoms. The magnification of quantities in our modern interdependent world, where the yields of a thousand harvests are aggregated into a single total and the slaughter of millions of cattle is treated as one statistic, staggers the imagination of those who think quite easily in hundreds of dollars, scores of people, a few acres of wheat or corn, and ten new babies this year. Or consider the sheer vastness of the ground that is covered so swiftly in an airplane, but so slowly if one tries, if only in imagination, to walk on the ground from one airport to another. The clocks in the airports tell us what time it is — in Tokyo or London or Tehran — but not whether it is day or night. We need new instruments, new ways of seeing, new ways of hearing, and new ways of thinking about the whole world at once.

Long ago, one of the first marvels that was achieved by the application of optics and the technology of glass manufacturing was the reducing glass, which was developed for the use of landscape artists and mural painters. Faced with the problem of rendering a wide landscape within the dimensions of a relatively small canvas, the artist could see in the reducing glass the scene diminished to a miniature; he had to look sharply to pick out details, yet the whole scene was still there. Just as we are able to reduce to meaningful comprehension the numbers in which we reckon people or tons of grain or dollars (or rubles or yen), through aggregating statistics from all over the world in tables and charts, which, for example, reduce billions to single numerals or dots, with the note "each figure stands for 1,000,000,000," so we now need a way of dealing with the total picture of the world, a way of both expanding our scope of vision and yet of sharpening the focus of our eyes and thought.

But in addition to such a macroscope, there is also a need for some way of including what we do not know with what we do. This is done at present by large, loose categories that make it possible for us to believe that we are thinking about whole civilizations or whole continents: the East and the West, the developing countries, the Third World — a phrase which is sometimes applied to countries and, sometimes, simply to people whose skin is not "white" — the poorest fourth, the oil-producing countries and the oil-importing countries, the literate and the illiterate or nonliterate.

When these simple, large categories are used, the extraordinary diversity and richness of the components are lost, as if, when a reducing glass was lifted in an artist's hand, the colors were to turn to different shades of gray. Thus, phrases like "primitive societies" and "traditional societies" actually kill imagery, because one must deal with a rush of unrelated images (like the airport posters in which the wonders of the world are set at angles to each other), or names and numbers must be substituted for any imagery at all, and these move in a blank procession through our heads, little boxes, little black boxes into which we never look. Or evaluative adjectives are substituted; thus Lyndon Johnson — who was an expert in the exact shade of humor which was safe to use in northern Georgia but not safe in southern Georgia — when he spoke about the rest of the world, resolved it into good countries and bad countries, which were together involved in a game with only two sides. The search for simple dichotomies may leave out a large part of the world altogether; the massive obsession of the United States and the Soviet Union with each other, each believing that the world was a battlefield for the two of them, resulted in China being omitted from their calculations for almost a decade.

And so the task of finding ways to reduce scale and bring dimensions to a manageable size — and yet retain the scene in all its vividness and concrete reality — is a very pressing one, pressing for those of us who learned to think so differently and pressing for the young who are struggling to obtain some sort of mastery over this mass of fact and fancy, spewed out by the many media which bring us very close — but somehow not close enough — to the world around us.

Different people have tried different ways and used different models to do this. Phyllis McGinley, whose easy, humorous characterizations of the great — ancient saints and contemporary prima donnas — delighted a generation, used the people of the small town where she had grown up, relating the famous or dead personalities to those in the tiny galaxy of the village green. People who have lived intensely in a small, well-known milieu find it easier to grasp a similar milieu than those who have grown up in the relative anonymity of a large city where only a few parts of the city are really known and the rest are names on a subway map or places visited once long ago to see a circus, or attend a funeral, or buy a special kind of spice. It is less of a shock for city people to move to a completely different large city in another country than it is for people who have lived all their lives on isolated farms or in small towns to move to a large city. And, those who have been members of a highly privileged class in one society find it easier if, when they travel or emigrate, they can join an equally privileged class in a different culture. And the very poor may migrate from one country to another, where they continue to live their lives in the same kind of small, meager niche, which is nonetheless understandable because it is similar to the kind of life they left.

In writing, an anthropologist uses another model, one that is not based on the realities of the time and place where he or she was born and bred, but instead one based on a purposeful and disciplined entrance into another society to learn what that society has to teach. To do this, even in the very simplest societies we know, is, of course, an act that involves an extreme degree of hubris. How can one ever hope to know the intricate responses of hunting peoples to each bent leaf or twig, or the delicate nuances of vocabulary of a people who care about differences in color so fine that they defeat our eyes and, so far, our cameras? Is not the attempt to put down on paper the whole culture of a people, whether they be the Bushmen of the Kalahari or the Pygmies of the Ituri Forest, and to order these details in monographs with headings such as "Social Organization," "Kinship," "The Yearly Round," "The Life Cycle," "Ceremonial Life," an impossible piece of reductionism?

This is a problem that anthropologists have been struggling with ever since we began to write serious scientific monographs about people among whom we have lived for the sole purpose of studying and recording their way of life. Every complex event has to be dissected into single aspects. The behavior of someone who is a chief, a leader, a scoundrel, a mother's brother, an eldest son, a man with magical powers, handsome and tall, thrice married, or childless has to be placed in a whole series of contexts as we discuss chieftainship, leadership, the ideas of right and wrong, kinship functions, primogeniture, sorcery, ideas of physical beauty, marriage, or the penalties of infertility. What is left of such a man, who is referred to in the catechism service as "N or M" and whom we call Smith or Jones, after his "aspects" have been neatly stored away in file drawers, someday to become chapters in books, and still later to be refiled in cross-cultural indexes under the headings of "Sorcery," "Kinship," "Primogeniture," etc.? Isn't this just what statesmen and planners do when they pigeonhole nations as "industrialized" or "unindustrialized" and "rich" or "poor"?

Stated by itself it is. And many times, this is as far as an ethnographic description goes, giving us only terms arranged in categories, only bare bones stacked in piles. But the good field-worker goes further. If one lives in a hamlet or a village and learns the relationships of each person to each other person and the place of each in the village, the whole remains meaningfully interrelated and each person, however many his or her aspects, remains a whole person. But as long as we had only words to use in describing them, words which had to be placed one after another in linear fashion, taking whole chapters to describe single events, it was hard to communicate the sense of wholeness which the field-worker had experienced. And it was hard to include, in the description of the behavior appropriate to a mother's brother, his costume, his personality, and his state of mind, as well as the image of his shoulders sagging with sorrow because he had just lost his third wife.

It was the introduction of the camera which changed all this. Now one could relate one aspect of posture, costume, gesture, or place in a ritual whole to another taken from some other event, without losing the total complexity which each individual embodied. The posture of an old man watching an airplane and of a young man carrying a small girl possessed by an angel could be compared without losing the lines on the old man's face or the slope of the shoulders of the young man who carried the little trance dancer.

Further, as we became more sophisticated, we began to learn how to include what we did not know in thinking about what we did know. The simple ignorances were easy; one might come away from a long stay in the field without having seen the death of an important man, or an initiation ceremony held only once in a decade, or a feast in a temple that was visited only every fifty years. From these simple recognitions we learned to specify that we knew something about the art but very little about the music, something about agricultural ceremonies but very little about the long, arduous hours in the planted gardens, something about how the people felt who feared sorcerers but very little about how the sorcerers felt about their victims. Each piece of partial knowledge became illuminated, glowing differently because of the recognized areas of greater darkness around it.

It is something like this that we have to do with the world at present if we want to think about it at all without reducing it to dull categories of climatic zone or economic level, technological development or political orientation, or simply the size and location of nation-states. The pictures in this book are a small selection from the thousands of pictures Ken Heyman took, out of the millions of others that he could have taken. Each preserves its integrity while relating to some aspect of the other pictures; and each suggests vast areas of which there are no pictures, either in this book or in the imaginations of those who will look at it or who will read the words that accompany the pictures. One photograph of plowing with oxen suggests not only the tractors that are replacing the oxen, but also plowing with mules, or horses, or water buffalo, or animals whose names have only been encountered in crossword puzzles. But if the picture of plowing in Bali, sharp and clear and known, can be illuminated by the knowledge that in many parts of the world people use other animals and plant other grains, the fact that there is only a picture of one kind of plowing lets us think about the whole without making a category called "plow animals" in which all animals, horses, mules, oxen, and water buffalo become equally gray and shapeless.

This is what we are trying to do. The original choice of what to photograph was Ken's, and the first choice of which photographs to include here was his. In response my reaction to his choices, about a fifth of the pictures were changed to sharpen some point I wanted to make, or to alter an emphasis which I did not share, or to change the book from a reflection of his mood alone, as he faced an urban world that seemed to be falling to pieces as well as the seemingly changeless primitive and peasant peoples who so delight the eye, to a book in which my perception of the world, seen from a number of other perspectives, could be included. This book contains no photographs from New Guinea, where I have spent so much time and learned so much, because Ken has never been there. It includes photographs from places I have never been as well as from places that Ken and I have seen together and about which we share some of the same knowledge. I hope that it will contribute, in all of its diversity, complexity of experience, and ignorance, to making a special kind of macroscope of the world that we now have to think about as one whole.

I

HIGHWAY INTERCHANGE, U.S.A.

THE DREAM
OF
TECHNOLOGICAL
SALVATION

U.S.A.

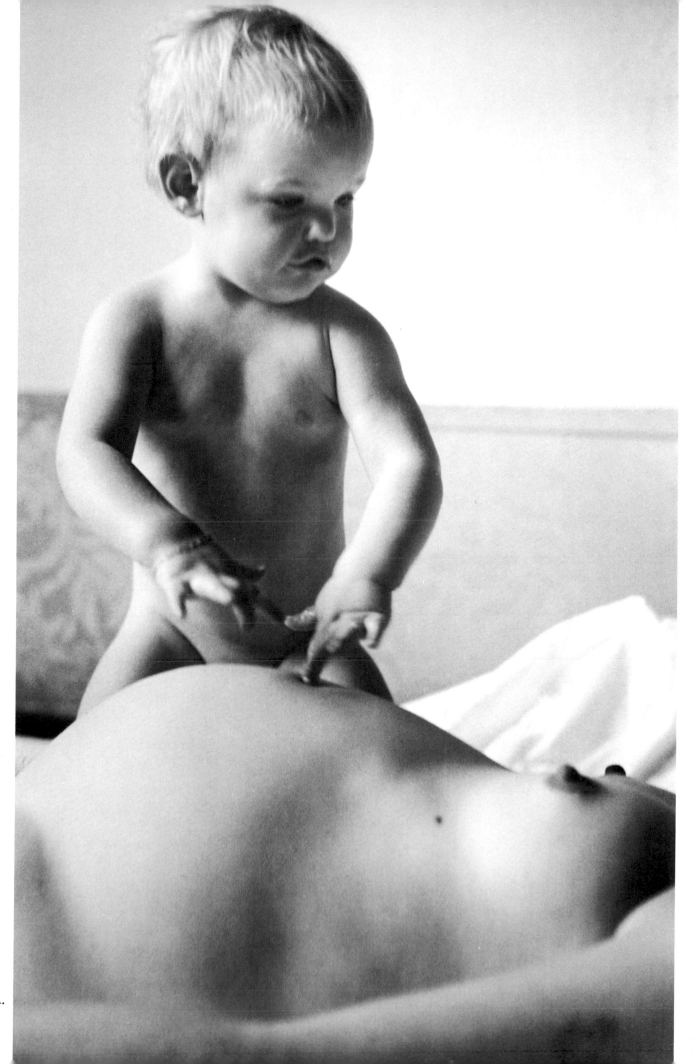

U.S.A.

DAY NURSERY, HUNGARY

WHEAT, U.S.A.

GYMNASTICS, FINLAND

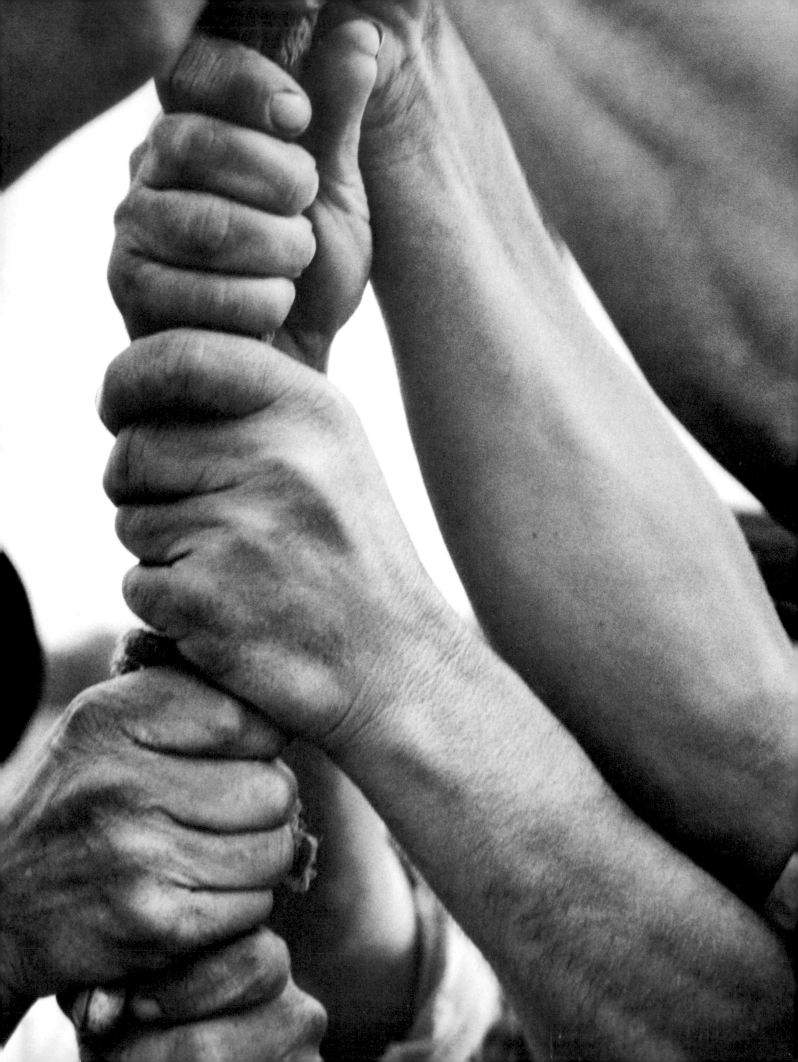

HANDS AT WORK, HUNGARY   HANDS IN DANCE, ISRAEL

HOUSING PROJECT, HONG KONG

SKYSCRAPERS, NEW YORK CITY

FAVELA, BRAZIL

HOMES, NORTHERN NIGERIA

BOARD MEETING, U.S.A.

STOCK EXCHANGE, PARIS
STOCK EXCHANGE, CALCUTTA

MECHANICAL DRAWING, U.S.A.

OIL WELLS, U.S.A.

WHEAT FARMER, U.S.A.

SPACE CAPSULE ASSEMBLY, U.S.A.

BICYCLIST, NIGERIA

NIGERIA

U.S.A.

PERU

NIGERIA

19

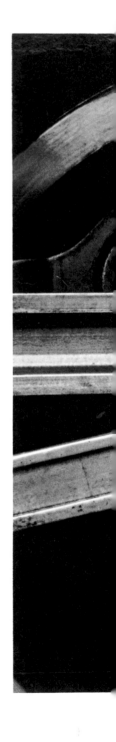

ENGINEMAN, U.S.A.

INTERSECTION, COPENHAGEN

GOLF RANGE, TOKYO

HARVESTING WHEAT, U.S.A.

24

PLAYING GAMES, LONDON

WAITING ROOM, U.S.A.

DOG-WALKER, TOKYO

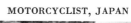

AMUSEMENT PARK, JAPAN

PREMIER, NORTHERN NIGERIA

ANTIWAR RALLY, NEW YORK CITY

ABOVE: CONFRONTATION, TOKYO

TOKYO

Shared
Hopes

Historians divide the passage of time since the beginning of written records into periods, but as the criteria are imprecise and confusing they never quite agree about the length or boundaries of any period. For each of us, the period in which we have lived as human beings conscious of the world around us is distinct from "history," which we regard as the time and the events that happened before we ourselves were born. As the future has become a subject of study, it has come to mean the way things will be and the things that will happen after we are dead. And, the more we know about the history of the universe, of our solar system, of the beginnings of life, and of the early history of our species, the more the present becomes an episode in a partly known and partly unknown process of change.

During my lifetime, archaeologists and paleontologists have pushed the history of humankind back millions of years, explorations by plane and rocket and satellite have pushed our knowledge of our planet and solar system back farther, and radio telescopes have pushed it to the outer limits of the cosmos. (Interestingly, *Pioneer 10* would be a UFO, an Unidentified Flying Object, if viewed from another solar system.) We have added the concept of inner space to refer to the world under the sea and to the interior of the earth. The parallel concept of "inner space" as the poorly explored inner life of human beings has also been invented. After the first fever of excitement over the exploits of the cosmonauts and astronauts, children soon turned from drawing pictures of *Sputnik* and outer space cluttered with discarded stages of rockets to drawing frogmen under the sea as well as scenes from the life of early man, which became more exciting to explore in the imagination than the stereotyped images of buffalo-hunting Indians. We can see here an attempt to arrive at some kind of balance between the eons behind us, before human beings were self-conscious enough to wonder about the stars, and the distances that it is now possible to venture with sensing instruments that report back to us.

The limits of imaginable life were pressing forward in time and out in space, as well as back in time and inward in space. At the same time, the extraordinary things that were happening in the world, as we began to speak familiarly of other worlds, came to be taken for granted. If John Glenn could look down on all the cities of the world, such a thing was no longer an event of supernatural significance as it once seemed when people read about the temptation in the wilderness, of the devil taking the fasting Jesus to a high peak where he could see all the cities of the world and offering him dominion over them. It ceased to be remarkable that we could watch events transpiring on the other side of the world, and small children no longer wondered how the Australians, way down under, managed to stand up, upside down.

Within this expansion of our universe, the events taking place on earth ceased to seem amazing, and people almost forgot that before World War II we had not lived in a known, mapped, and interconnected world where it was commonplace to find, in any large hotel, men and women, all wearing the same uniform dress of the industrial world, but carrying in their faces, in the color of their skins, in the texture of their hair, the marks of tens of thousands of years of isolation. Now they come by plane, as once they came on foot, on camels, and on horses, to the old marketplaces and centers of political power, to Persepolis, to Delphi, to Peking, and to the great cities of Yucatan and Peru in the New World, which were cities long before the peoples of the Old World knew that the Americas existed. The life styles of Euro-America have spread so rapidly, so little challenged by the diverse peoples whose own ways of life were affected, that it is almost impossible for the middle-aged to realize that when prophets speak today of the beginning of a new era, they are referring to a system of interrelationships which are only a quarter of a century old, a world style that has been hammered out with unthinking speed and carelessness from the ruins and the hopes that emerged from World War II.

While the outcome of that war was still uncertain, the foundations of our era were being laid down: the development of the atomic bomb, of computers, of rockets (that later launched satellites), of television, and of automated factories; the growth of the expectation that science-based technology could solve any problem, social or physical, and that both war-making and peacekeeping, fighter planes and freight planes, manufacturing and agriculture, could be rationalized, programmed, and made progressively independent of human hands and human intelligence. The treatment of human beings as a part of a machine process, although as links that we could not yet do without, was associated with a hope that the new high technologies would produce plenty for all, a world free of hunger and want, where a literate, healthy populace with fewer and fewer hours of work a day could enjoy everything that made life worth living.

These expectations of salvation by science and technology were accompanied by a set of assumptions that were not seen — at least, for a long time — as having direct relation to them. Thus, it was assumed that the appropriate unit for political decision making was the nation-state, with the unspoken proviso that small and weak nation-states would be convenient units for economic exploitation by the largest nation-states under the guise of international trade or a shared ideological commitment. And it was assumed that human dignity as a goal of organized society was firmly associated with political self-determination, on the one hand, and material well-being, on the other. The tension between the two versions of this ethic, the Soviet and what came to be called by its members the Free World (and by its critics the Capitalist World), was expected to speed up progress all over the globe. According to the accepted theory, as these two groups of nations with their contrasting views of the world vied with each other for the dependence, resources, and trade of the developing countries, all would benefit.

There was so much emphasis upon the contrast in ideology, fueled by political rivalry for world dominance, that those parts of the world that were the recipients of the propaganda and favors from each side hardly noticed how very similar were the basic assumptions of contemporary Capitalist and Communist ideology. Both believed in bigness, centralization, and industrialization; both trusted high technology, both believed that the proper goals for human beings were the possession of material well-being and the exercise of cherished rights — although the Soviets emphasized economic freedom and economic security and the Free World emphasized political freedom and individual autonomy. Neither group relied upon religious hopes for a better life after death to excuse the planners of the present world from their failures to arrange a better life on earth. Both pursued an arms buildup that was more competitive than functional so that it became possible to construct whole systems that shortly became obsolete as the other side introduced technical improvements and to pile bomb upon bomb in a numbers game in which each had enough striking power to destroy all of humankind many times over. With this process, both found themselves with enormous stockpiles of obsolete equipment of every sort, which it then became convenient to unload in the secondhand arms markets of the rest of the world. At the same time, both proclaimed as official doctrine the desire for a just and peaceful world, and each presented its own system as the only route to attain this goal.

Furthermore, the victors and the vanquished in World War II, whether united against Nazism or seduced into its destructive values about race and blood and soil — ideas that conflicted with both Christianity and Marxism — came out of that war ready to set to work vigorously to rebuild their cities — from Tokyo to London and from Moscow to New York City — and their economies in the same image.

And, they shared the view that the old forms of colonization were obsolete and that every people who had been living under the political hegemony of the industrial world should become free and autonomous nation-states, given dignity by their equality with all other nation-states. Groups of old nations, however small and economically and politically dependent, were to become more tightly bound to one another, through the pacts which united the countries of Eastern and Western Europe, respectively, and the attempts to build more lasting alliances among the nations of Latin America.

The image that was used most often was that the economically less-developed countries, progressively freed from their former colonial subjection, would reach a "takeoff" point and "catch up." However much the Russians and the Americans might quarrel as to which was the First World and which the Second World, they both agreed that there was a Third World, containing most of the world's peoples, who were not industrialized and who lived in a style which provided, at most, a bare subsistence rather than the affluence that was the dream of the industrialized world, whether actually achieved or not.

By insisting on the termination of colonial imperialism and the overriding importance of political autonomy, both sides were able to participate in the destruction of their own old orders and the establishment of basically similar forms of modernization. The leaders of the Third World also came to share the goals of industrialization: high technology, literacy to be achieved in a school system modeled after those in the older industrialized countries, heated and air-conditioned buildings for work and residence, motor transport, television, commercialized agriculture, a nationally owned airline, and a favorable balance of trade. The sheer nonsense of each country hoping for a favorable balance of trade — beautifully caricatured in the suggestion that we all export to the penguins — was unrecognized by the planners caught in an outworn competitive ethic.

Implicit in the expectations of every country, as each and all — whatever their previous cultural and religious beliefs — became involved in this dream of modernization, was the belief that every economy could grow and grow and grow, that it could make ever more things, create ever new demands, feed ever more people, and find ever new ways to use the resources of the earth, happily and easily providing substitutes when any particular resource ran out or was inconveniently expropriated. The bigger the factories, the bigger the fields of grain, the bigger the machinery, the better off the common man would be. In the Soviet Union, the bare minimum was distributed more equitably; in the industrialized Capitalist societies, the share the poor were expected to get was more impressive: more cars, more gadgets, more variety in food and clothing, more freedom of movement.

The devastated countries of Europe, defeated Germany and Italy, victorious Britain, France, and Russia — and conquered Japan — began rebuilding their cities and spreading out over the world, while the United States, its cities undamaged by the war, participated in the same set of dreams: big agriculture, big factories, big unions, and big government. Each of these countries had a tie to the nations of the developing world, all of which were in need of technical assistance so that they could reshape their primitive and archaic political and economic systems to be like those of the industrialized countries. Back of the whole set of premises lay a naïve kind of evolutionism which assumed that the stage the modern world had reached was higher, better, and more economically sound than any that human beings had reached before.

The association between technological solutions and the ideas of bigness and growth, which came to be viewed as almost absolute goods, was accompanied by a number of important inventions, including oral contraceptives and the "miracle grains." It was expected that population could be controlled and people fed by these means, without any deep involvement or changes in the human cultures affected. The billion new mouths that were appearing on the scene could be fed by "miracle wheat" and "miracle rice," new species of grains, that, given enough pesticides, fertilizers, and water, would provide many times the yield of traditional grains. The new methods for controlling population and increasing harvest yields seemed so overwhelmingly superior to the old, clumsy methods that the voices of those who worried about the danger of a single crop that could yield to devastating pest or drought or flood were drowned out by the enthusiasts.

The acceptance of these new technologies in the developing world was powered by the strongest motives that can move human beings: a concern for the well-being of those with whom they identify, whether as members of the same family, the same clan, the same village, the same race, or the same nation, and a desire to reduce the denigration of one's own group compared with members of other groups in the world. Of course, this rapid spread of a single-minded belief in the total goodness of economic growth toward universal affluence also served the purpose of power-hungry bureaucracies and profit-hungry corporations. Indeed, they became part of the push toward endless growth, caught in the same trap as those with the lowest economic conditions and the most extravagant hopes. And, the larger the enterprise or multinational corporation, the bigger the country, the more the search for power came to override considerations of prudence and even profit.

As this new life style battened upon immense natural resources imported from far away, the extensive division of labor, and the development of great distances between producer and consumer and between home and work, the older systems, in which people grew their food and purchased only a small part of what they needed to live, seemed old-fashioned, inappropriate, "irrelevent," something for those educated in the new way to feel ashamed of. And such discrepancies led to proliferating hypocrisies: when the Soviet bureaucracy used to demand that all the badly made and out-of-repair tractors start the spring plowing on the same day, the peasants on the farms did all that could be done — they got them in line and moving, if only for fifteen minutes — so that one more stage in a five-year plan could be reported; but in the end there was too little bread. In the industrialized countries, old buildings were torn down, and newer, shoddier, hastily designed, internally inconsistent, bigger buildings took their place.

Workers drawn from populations who had worked almost naked, or in sarong or kilt, now wore pants on the streets of large cities because they thought that wearing European dress would increase their dignity. But the pants were often old and dreary, and the clothing that originated in Europe fitted strangely in the new scenes. Critics of the slow rate with which the developing world was "catching up" or of the methods by which the developed world was helping them to "catch up" (lending them capital which would involve them in hopeless debt) were also critical when any act that was performed by machines in one place was still performed by human hands somewhere else.

And yet the incompatibility of the old with the new, of the small with the large, so often seemed irresolvable. Planners looked at the strange distribution of agricultural land in ancient villages — a plot here and a plot there, one in the lowlands, one in the hills, sown with seeds so heterogeneous that a plot of the same grain looked more like a patch of weeds than serious farming — and advised merging the scattered plots into large holdings that could be plowed and harvested by machines. Soon, ten small farmers, who could survive somehow because some of their hodgepodge of grain was drought resistant and some flourished when there was too much rain, were displaced by the one man who could accumulate enough capital to buy or rent the big machines. It seemed so clear to those who came from the wheat fields of Europe and the Americas that big holdings, which were more economical once planted in miracle grains, were required if their promised tenfold yield was to be achieved.

It was also held that the farmers who lost their land and jobs in this process could simply go to the cities, where new industry would give them work. And, when these rural people arrived in the cities, housing could be constructed with modern, efficient methods which would be so much more hygienic than the insect-infested huts they had left. People who had given up building stone, or brick, or even thatch, houses, each detail of which had been beautifully fitted together by the generations who had lived in them before, could be given apartments of precast concrete. And, in the new economics, the houses that people built for themselves became too expensive; the cost of a collective feast for the whole village who helped to raise a house became too great. As the population grew, finding thatch within a feasible distance became harder; sheet iron was — everyone pointed out — actually more economical because it lasted much longer.

And the people who were leaving their houses of clay and bamboo agreed. The postwar world was not a world in which the standards of the industrialized were thrust upon the rest, as so many critics like to think, bracketing the missionaries of old as equally to blame with the technicians of today who believe that large-scale manufacture at a distance, based on a rapacious exploitation of nonreplaceable and replaceable resources alike, would provide a better way of life. It is often forgotten that the missionaries who traveled to faraway islands, because they believed that their Faith was the only faith that could save the souls of those who were living in outer darkness, also brought with them products of Western technology: scissors, compasses, and fountain pens, wire for fishhooks and steel axes, woven cloth that could be washed many times and soap to wash it with, steel needles and sewing machines to make it into clothes, pots and pans that did not crack and leak like those made of clay, and iron pipes, hammers, saws, and nails. For example, missionaries came to the Pacific islanders I have studied, in ships, larger than any the island peoples had ever seen and built to sail safely on more turbulent seas. All of these things, the missionaries knew, were better than the things the islanders used, just as they knew that their Faith was superior to the belief in spirits that lurked in the local waterfalls. Their belief in what was good was all of a piece, and they did not distinguish between the hope of heaven and the technology they brought.

There was much in this that the Pacific islander could not deny; the wire did make better fishhooks, steel tools were far superior to stone, and some work could be shortened from a week to an hour, even with the preindustrial technology the newcomers offered together with a God under whom all men were brothers — even the people on the next hill against whom they had carried out headhunting vendettas for thousands of years. And for the regulation of life by threats of sorcery and threats of personal combat with spears or clubs, the newcomers brought a majestic concept called the Law, before which all men were equal. However shoddy and exploitive the administration of that law, however petty and inconsistent the embodiments of the Christian tradition might be, the vision they brought was shining, the poor trappings irrelevant, like the explanation given children of sacraments: however unworthy the celebrant, God is present in the host, just as gold coins are still gold in even the cheapest and shabbiest purse.

So the first missionaries and the first conquerors brought horses and guns to American Indians, and steel and cloth to the peoples of the Pacific islands. And later, after World War II, they brought the islanders marvelous machines, outboard motors that shortened a week's paddling upstream to half a day, flashlights and cigarette lighters that stayed alight when leaf torches would have been quenched by rain, road-building machinery to uproot trees and lay cement roads, new seeds for new crops, and the promise of overseas markets for cocoa and coffee and palm oil, properly processed by machinery imported from Europe.

Those who brought the new wonders and those who received them shared the feeling of entering a better and greater era. This has been the way in which the discoveries of one civilization have spread to others from the very beginning. One human culture learned from another that with the bow and arrow they could substitute death at a distance for the risk of death with a club; they learned to make fire and not just to catch and keep it; and they learned to fasten a steadying outrigger on a dugout canoe for ocean travel and to remove the outriggers for travel up a river. The advantages of technical inventions are easy to grasp, and when human beings have known the amount of energy required to fell a tree, or transport a log, or hollow it out to make a canoe, the appreciation of toolmaking and tool using is immediate. It doesn't require previous experience with earlier forms of transport, whether on foot, by canoe, by dogsled, by llama, or by donkey, to appreciate the speed and carrying capacity of an airplane, although people who have only seen airplanes in the sky may ask whether one can eat and sleep and defecate inside one.

Furthermore, when schools and colleges were established, they imitated models from the industrialized countries where the first new leaders were educated, and the value system which finds pleasure in city skyscrapers, linked by three-level interchanges, became the one that was taught to all those educated in these schoolrooms. The greater the pride in being now a citizen of a nation-state that is coequal with all other nation-states, whose ambassador sits in the United Nations (its name in letters the same size as those for France, the USSR, Brazil, the United States, and the People's Republic of China), the greater the eagerness of the newcomers to the modern world to match the possessions, the constructions, the trade, and the large-scale enterprise which the "missionaries" from that modern world had brought to them.

Ideologies become subordinated to the pride of nationhood: in Romania, the most important consideration is that the people are now governed by Romanians. In convents in Romania, nuns now make traditional religious vestments using electrically driven equipment. There are factories which make tractors, in various shapes and sizes. The shiny presentation catalogues of the factories, which might have been located in Tokyo or Detroit, Birmingham or Bordeaux, are printed with the most exquisite color rendering. The tractors go all over the world, no more adapted to the needs of the people who will use them than those that come from any other factory; the recipients must adjust to them. So a center from which modern machinery goes out becomes a kind of diffusing center; a factory located in a country like Romania with a very old, literate high culture, may send out machinery to India, where human beings tolerate incredible deprivation in the hope of a better life in the next incarnation or in the fear of a worse one, or to the Pacific island of New Caledonia, where an urban center is being built with skyscrapers on terribly expensive urban land, hundreds of miles from anywhere.

Mass manufacture, large-scale production, and promiscuous and miscellaneous marketing without regard to the distance from the point of manufacture or the suitability for the climate and the habits of people assumes the same form whether managed by Europeans or Japanese, Pakistanis or Canadians. The delicate calculations, carried out for hundreds of generations, of which field to plant this year, which seeds to save, how to pick enough reeds to weave fish traps without depleting the supply of reeds, what to plant for one's children and what for one's grandchildren, have all been torn apart, everywhere in the world, as we plunged headlong into a mining, building, transporting, inventing binge of unprecedented proportions, taking from every part of the globe and sending to every part of the globe, distributing everywhere examples of the wonders of Euro-American technology. The skyscrapers with their electrically controlled temperature make it possible for the light-skinned to live comfortably in the tropics and the dark-skinned to keep warm in the Arctic, reassuring us that we can even build livable structures in outer space.

Only as those among whom this high technology originated begin to question whether human beings are becoming, indeed, just "human components" in an interlocked global system so fragile that one small country could disrupt it, begin to wonder whether we have not, somewhere, taken a wrong turning, grown too big and too fast, will the peoples who have received these products of technology last begin to question their value also. For they see themselves, much as we have seen them, as people deprived of the technological wonders of modern life, people waiting for their fair share of the same world. So an ironic twist was given to the United Nations Conference on the Human Environment in Stockholm in 1972, when, after Barbara Ward had proposed that we give just "one percent of the gross national product of the developed world to help the developing world, with the fear the Lord might think it was not enough," the representatives of developing nations were reported to have said: "Give us just one percent of your gross national pollution and we'll be satisfied."

2

The
City
of Man

As the world became more urban, the different ways in which cities were conceived became more clearly defined. The dichotomy between urban and rural has been a part of human thinking for several thousand years, and the distinctions that are made today began long ago in folktales; it has continued in the morality preached by new sects promising salvation from the sins of the city, in the adjectives which are used to describe the city dweller and his country cousins, and in the aims and hopes that guide peoples' aspirations. Is the city the epitome of civilization or is it its downfall? What will come after urbanization in the different parts of the world? Are cities properly to be seen as places where the rich and powerful gather around them a thousand slaves to do their bidding, or a million landless poor to serve their needs, or a voiceless, miserable, urban proletariat to follow their orders? Are these urban masses happier than the contented, honest, hardworking peasant farmers who are safer from ruin, whose wants are fewer and easier to satisfy, and whose tragedies are more manageable because they are seen as acts of God?

One way of thinking about cities is to assimilate their growth to the whole process of human evolution. In such a perspective, we can see human beings as creatures who have evolved through the extension and work of their minds and hands and feet and backs, gradually encasing themselves in a shell of their own construction, very much as the snail encases itself in the shell which it has constructed from its own physiological resources. Thus, through long ages human beings have been learning to live more comfortably, to acquire a more reliable food supply, warmer, drier abodes, clothing to protect their skins and their vanity, containers in which possessions could be transported and infants protected from prying eyes. During these millennia, simple technologies became complex ones, the division of labor increased and specialists emerged, the capacity to organize larger and larger projects, such as the construction of temples, dams, aqueducts, and bridges, and to move farther and faster by the help of horses, camels, and ships with sails or many oarsmen, developed. All of this could be seen as underpinning the inevitable and natural growth of other parts of the culture we have come to call civilization: the provision to a few of the leisure to pursue philosophical speculation, to develop mathematics, to produce paintings, sculpture, and majestic buildings, to compose and perform music and plays, indeed, to originate new schools and styles of art, to develop laws of justice and rules for commerce, to pursue religion and liturgy, in sum, to achieve grandeur of spirit and grandeur of life.

The image of the snail ever producing a larger shell has been perpetuated in the imagery of "The Chambered Nautilus." The poet exhorted himself:

> *Build thee more stately mansions, O my soul,*
> *As the swift seasons roll!*
> *Leave thy low-vaulted past!*
> *Let each new temple, nobler than the last,*
> *Shut thee from heaven with a dome more vast,*
> *Till thou at length art free,*
> *Leaving thine outgrown shell by life's unresting sea!*

Whether the image of the city, with its many "mansions," is seen as a religious symbol of the hope for a better afterlife, or whether the city is seen as a peak in human experience one attains only to find it necessary to climb down on the other side and return to simpler environs, "where neither moth and rust doth corrupt, and where thieves do not break through nor steal," these differing conceptions lie back of the way cities have been built, of how they have been regarded by those who dwelt there and by those who aspired to dwell there. Cities have been seen both as an enhancement and as a stumbling block to a fuller human existence.

In modern thinking, the tremendous increase in the speed with which we can move has influenced in many ways the manner in which we think of cities. One of the most important has been the breakdown of the rigid dichotomies between the city and the countryside, as involving two distinct ways of life. Whether the contrast was between the nature of the people who lived there, or between different degrees of population density which required different kinds of sanitation, or between the natural and the unnatural, these dichotomies have dominated our thinking. The city was good and the country was uncouth and rustic, the abode of ignorance, superstition, and distrust of progress; or the city was the destroyer of men's virtues and corrupted their souls and the countryside was innocent and good. For a very long time, perhaps for as long as human beings have had the means to build a significant city, there have been recurrent worries about the relative goodness of a concentrated or dispersed way of life.

The process of urbanization can be seen as a kind of rhythmic historical process in which certain societies invented forms of social organization which first made it possible for people to meet occasionally, then to live part of the time, and finally to live all their lives in huge urban aggregations which were dependent upon a distant and different way of life for their basic necessities. Often, however, these cities became too large and too top-heavy; in times of difficult economic circumstances, they could no longer pay their way. Then, the surrounding hillsides are denuded, or the avariciousness of those who have no such cities is aroused and cities fall to conquerors; centuries later, peasants plowing their fields turn up the broken shards of former grandeur, and scholars, themselves the product of cities, come to excavate the glory that once was, to moralize and speculate about what evil brought their decline, what historical forces conspired to bring about their fall.

But this kind of thinking has its dangers. Cities of many kinds — pilgrimage cities and cathedral cities, trading cities, cities formed around coliseums and amphitheaters, seats of government and power, and forts and seaports — all get muddled together in the attempt to derive the concept of an inevitable process of urban development that has gone on and will continue to go on. Such a theory must explain a wide variety of urban phenomena. Consider the tendency of cities all over the world to model their public buildings after those of ancient cities: the courthouses and capitols of provincial states have Roman facades, while a church in New York City imitates the mosaics of the Middle East. The architecture of an earlier period later comes to express public feelings of pride and glory. Thus, nineteenth-century railway stations borrowed the trappings of Roman power. Or consider the efforts that city dwellers have made to import into the city the delights of the abandoned countryside: parks and pools and artificial lakes, bosky dells for forests, space recovered from overcrowded inner cities, gardens (where flowers can be viewed but never plucked), bridle paths and bicycle paths and skating rinks and swimming pools, restoring to the city's unnatural aggregation the beauties and pleasures of the forest, the seaside, and the countryside.

One of the characteristics of this restless century that is drawing into its last quarter has been the ongoing discussion about cities: Are they a natural accompaniment of the evolution of humanity? Do they follow laws of growth and decay, as living organisms do? Are they to be regarded, at an ecological level, as providers of more and more niches within which different species of humankind — thinkers, artists, engineers, printers, lawyers, doctors, and workers of every variety — can live in ever-growing interdependence and productivity? Are they a kind of climax phenomena, similar to growths of vegetation which, ecologists note, can attain an equilibrium so satisfactory that a stand can last for centuries? Or are cities perhaps a very dangerous phase in the evolution of the species, which may result in the destruction of a whole way of life or life itself?

And the arguments go on. Meanwhile, and especially in this century, the city has become the manifestation of all the rewards that our system of the extensive division of labor and interdependence can give us. Many have come to see the city as a place where everything one wants is within easy reach, where one is only forty-five minutes away from anything one might need: a physician, a library, a shop with all sorts of goods, people with whom one might want to talk, a play or film one wants to see, an opera or concert one wants to hear, a lecture where some new discovery is being discussed, a meeting of one's peers which is large enough to be challenging and command respect, a place where there is something for every age: a circus for children, a discotheque for the young, diversity and stimulation for the middle-aged, tranquil parks for the old, who may say as my mother's mother said the summer she was ninety, "I don't think I need to go to the circus every summer, do you?"

A city is a place where there is no need to wait for next week to get the answer to a question, to taste the food of any country, to find new voices to listen to and familiar ones to listen to again. It is that place where one need never be bored, where there is always the possibility of a new encounter that may change one's life. A city is a place where horizons expand, where one's possibilities are not shrunken and limited to the low expectations that some-one, kin or neighbors or the inhabitants of a nearby town, had for all the members of one's family line. It is that place where friendships are matters of choice, where love is always just around the corner, where the shortest walk can be an experience of surprises and delight. A city has a soul, while small towns, which lack the accompaniments of high culture, can be relatively soulless, given to too much suffocating intimacy, with no surprises, and a day-by-day routine life without excitement or promise.

As we built new capital cities, such an Canberra, Ankara, and Brasília, as older cities, such as Athens, Accra, Delhi, Mexico City, Singapore, and New York City, grew, this glowing image of the city survived to inform enthusiasm and to comfort those who saw only evil in urban poverty and urban blight and the growing number of shantytowns. Many continued to believe that these properties of the good life could be found only in cities and were essential if human life was to continue to develop, if the arts were to flourish, science progress, and the marvels of modern medicine be made quickly available to the stricken.

This powerful dream of the city triumphant continued to override the harsh reality that for most city dwellers the good life was still a dream. In many cities, they could not move easily from place to place — distances were too great to walk, walking safely was often impossible, and there was no public transportation; often there was no welcome for those who came from another part of the city, even to spend their money. They had not learned to value the entertainment that the city offered; they had few tastes that it could gratify beyond those of joining tremendous crowds at bullfights, or at football games, prizefights, parades, or riots. It did, perhaps, present an alternative to rural poverty and starvation, but what else for so many millions who came to the cities of the world?

And yet, just as it seemed that the cities would drain the lifeblood out of the countryside, it became apparent that the old division between the rural and the urban was crumbling, that one could live in at least a relatively rustic setting and work in the city, or live in the city and work in the country — modern transportation made either possible.

One could think of the city as a fortress against the inroads of a savage wilderness, or of the wilderness as needing protection from the savagery of the city. As vegetation once strangled the stones of ancient buildings, so cement overlaying the soil turns a city, however well planted with trees or furnished with masterpieces, into a depressing center of pollution and disease and crime. People with the habits of those who had always had miles of outdoors (in which their wastes disappeared under rain and snow) poured into the cities overnight, swamping every facility for safety and sanitation. Urban dwellers were then forced to travel farther and farther to find a bit of water in which to bathe or a green spot on which to eat a picnic lunch.

It was a modern Greek, Constantinos A. Doxiadis, who, in the presence of the glory of an ancient city that had been transformed by modern technology, and thinking of the sprawling cities of the developing world, came to the concept of human settlements as a continuum, to the notion of radiating circles of settlement, of zones of transition from primitive and rural to increasing urbanization, of water and light installations preceding urban dwellings. Doxiadis, building upon what was good in cities, upon what could be recaptured and expressed in new ways in a network of existing cities, developed the idea of a new form of urbanization: anthropopolis, or what Gottmann called megalopolis.

The controversy about cities has also drawn on the kind of rhythm of contrasts that has been so active in the formation of great human cultures, as each thing that is new and strange both attracts and repels. Once, rich city dwellers envied the poorest countryman — at the height of the envy, the English court went skating on the rarely frozen Thames — and rich men have assumed the habiliments of wandering beggars and hermits. Meanwhile, the countryman, hungry for adventure, dared the unknown terrors of the city, bound himself into years of indentured near-slavery, doing the most menial work, to taste the delights and the bright lights of the city. City-bred people became landscape painters, and country-bred people shut themselves up in laboratories to unravel the mysteries of the natural world that they had once had all around them.

With the kind of expansiveness that can be born of a sense of discovery, of open skies, and of incalculable technological advances, people began to hope that daily life would embrace both worlds. Radio, color television, inexpensive travel, paperback books, traveling clinics, the telephone, consolidated schools, all would bring to those who lived in the country many of the advantages of the city, along with good, year-round roads, electric lights, temperature control — and immediate access to the starry heavens. There were the dreams of what Doxiadis calls a Daily Urban System, larger than any proposed or existing megalopolis, with all real necessities (like a loaf of bread) close at hand and other less frequently used requirements farther away, and with connections between home and work, laboratory, shop, or factory, so that people could live in one place but reach an office eighty miles away quickly, so that the various members of a household could easily work in different places.

In this proposal, the division between city and country, between the fisher-folk who catch the fish and the urban population who eat it, between the ways of the city and the ways of the country, can be blended into a harmonious whole, farms and market gardens and chicken farms enclosed within the boundaries of a great and extended megalopolis.

These dreams must be considered as we look at what has happened to a world that became interdependent only within the last quarter of a century, a world that was suddenly able to undertake enormous enterprises, to build or rebuild whole cities almost overnight, to plan great industrial complexes, to turn the plains of the United States, Canada, and Australia into outdoor food factories, to hold the temperature within buildings steady during winter and summer, to design and build collapsible domes that could be packed in the rear of an airplane, then taken out, unfolded, and turned into a hangar; it was a world that could design structures that would shelter men under the sea as well as in outer space.

Back of these expansive dreams lay an age-old appreciation of the city as the center of civilization, the cynosure of every eye. For those nourished on the ideal of Athens, it was impossible to believe that from any city, any large aggregation of diverse people and talents, good things would not come. If the city were well planned, with open space, human scale, easy access, it would bloom. Doxiadis made a plan for revitalizing Detroit which assumed that those who had left the city loved it enough to want to return — sadly, as it turned out, while this was true of the residents of old-style cities in old countries, it was not true of the former residents of Detroit, or of many newer cities like Detroit.

When the idea of the city as the center of our most humane activities was added to the dream of the twentieth century as the century of the common man and of the democratizing of benefits, when the possibilities of rapid transportation based on the automobile were added to the availability to the many of everything that had once belonged to the few, the combination was almost irresistible. Despite the fact — as Doxiadis used to remind a group he invited each year to spend a week together on a cruise to Delos, the treasury of the ancient world — that builders are practical men who must start building tomorrow, especially if the enormous task before humankind of simply housing the burgeoning billions is to be accomplished, the fascination of this dream remained. To build a city, a real city, which would contain and express all the best that had been thought and said, painted and sculpted, written and sung, and yet a city in which all of these things would be available to everyone, not only to the rich and privileged — this was an irresistible dream.

And it is a very old one, a dream expressed by the story of the Tower of Babel, by the great ziggurat of ancient Persia, and by the jungle-strangled ruins of Central America and Cambodia. Humankind — the toolmakers, the house builders, the inventors of sledge and raft and cart, the inventors of the arch and the Gothic tower and the steel skyscraper, the developers of modern science on which our great technologies have been built — could do all that we had done in the past and, with the help of modern machines, of modern chemical processes, and of the modern understanding of the way systems worked, much more. We could do it without the costs on which every great city of the past had been built, without slavery, without serfdom, without indentured labor a thousand miles from home, without the immigrant lured from his pittance in some poor foreign village to starve in the cities, without the sale of indulgences and the corruption of the church, and without the cost in lives (as tunnels crumbled and bridges broke). The City of Man could be realized, with all the virtues of the past, now democratized and enjoyed by everyone.

Sometimes these goals were sought by abstractly nationalizing the palaces of the past. We then can have the strange experience of seeing thousands of quiet, joyless people walk with empty hands up and down the landscaped avenues of the Peterhof and peer through the windows into the empty rooms where Peter the Great once kept his treasures. Elsewhere, we see it in the plush student buildings of community colleges in the United States. We see it in the grandiose hotels all over the world which demand only money for admittance, regardless of race or social estate. We see it in the great television spectacles like Kenneth Clark's discussions of the history of art or Jacob Bronowski's presentation of the ascent of man, put on for everyone to enjoy, or in the practice of putting great orchestral performances on tape that can be played on stereos or over radios in village pubs. We see it in the burgeoning of universities and colleges all over the world. All of these were earnests that the kinds of culture that had depended upon the city for their existence could be democratized, could be made available to everyone, in the new affluence.

This dream was as sustaining in the first quarter-century of planetary interdependence as was the dream of the right of every human being hitherto poor, exploited, or excluded to share in the ever-growing wealth of the earth.

The two dreams met, sometimes with heartbreaking effects, but to understand what is happening in the world, in the second quarter-century of our planetary interdependence, the way that they sustain each other should not be underestimated. As culture has flourished and new inventions have spread and new civilizations have developed from contacts and exchanges in port cities, where the sea-lanes crossed and strangers met and talked, so also change and new hopes occur where dreams meet and feed upon and sustain each other.

3

Human Scale

One of the most puzzling concepts that modern city planning and architecture has introduced in the effort to correct for some of the inhuman aspects of modern building and mass production is the idea of human scale. Sometimes the phrase has been used to call attention to the fact that the whole scale of an object has been adjusted to the dimensions of a human body, as nursery furniture is adjusted to the young child or the steps of a building are adjusted to the stride of those who are expected to enter it. But at other times, in order to include under its honorific rubric architectural masterpieces — which have been the pride and focus of whole civilizations — the temples of Delphi, the Parthenon, and the great cathedrals have been regarded as, somehow, manifestations of human scale. The rationale for this was that they related, not to the dimensions of the human frame, but to the scope of the human spirit.

If we do try to see Notre Dame, Amiens, and Winchester, and Chichén Itzá, Persepolis, and Angkor Wat, as expressing "human scale," then the word "human" must be understood in terms of the meaning of humanity within each of the societies which built these great monuments, and in terms of the way any monument defines the glories of conquest, the mechanics of tyranny, or the worship of gods who transcend simplistic body-based values.

If, on the other hand, we think in the first set of terms, of a direct relationship between a chair or a table and a human body, we can end up with a technocratic daydream of comfort in which a guest, upon entering a home, mutters a few numbers that represent his height, weight, and somatotype, and his hostess pushes some buttons which order a computer-controlled, electrohydraulically adjusted "machine for sitting" to form itself to the scale of the impending sitter.

But a better way in which human scale can be introduced into our thinking is to consider a building or cluster of buildings as human to the extent that they take into account the needs of those who are going to live there, not only the universal needs for food and water and rest and excretion, but also the needs of the infant, the crippled, the blind, and the tottering steps of the aged.

So a building designed as a retirement home can be seen as human if it includes shallow steps, slanting stairways, and landings with seats, where the old who are frail, but wish to remain active, can pause to rest. A museum can show a welcome to children if there are ample hooks on which small children can hang their coats, toilets that are well marked and easy to find, water fountains a child can reach, and cases that are labeled at the level of a child's eyes. Here, attention to human scale means that the designer planning the physical arrangements took thought for those who would use the building.

Thus, human scale refers, in the first instance, to the intention of the architect or designer, to his or her sense of the needs of those for whom the building is to be built, and to the extent to which those for whom it is made are respected. Human scale is expressed in shoes for children that are as carefully made as the shoes of adults so that their lesser price would be a function only of the lesser amount of work and materials required. Another expression of this sense of human scale might be manifested by the designing of gym suits or of costumes for a school pageant so that the garments for the children are either inexpensive enough for every family to afford or durable enough to be used and then resold without placing a burden on poorer families.

Communities with human scale would be ones which meet the needs of small children for short, easily traversable paths, of cooks for nearby grocery shops, of children for a place to play within the sight and sound of the parents who are going about their domestic chores. But then as soon as we identify a particular thing as expressing human scale, questions arise about alternatives. Is it not better for every child to have a bicycle and for a community to have bicycle paths on which automobile traffic is forbidden than to restrict life within walking distance? Isn't this human scale too? And what if it is deemed better to have children gathered together in larger groups, where a number of teachers can provide for a variety of children's interests and capabilities. If this version of human scale is considered, then new problems confront us. The children must be collected in a bus to be driven to a central school. Then, what about the arrangements inside the bus? What about the size of the seats, the provision of windows that allow vision and ventilation (but not protruding hands and arms), mirrors in which adult drivers can keep an eye on the children, and racks for schoolbooks? And, if groups of small children are to have their view of the world widened by being taken on visits to airports, zoos, and factories, where there is a danger of their being lost, should they be fastened together with a long cord or taught to march in strict lines? Or, should there be a lost child center in every public place, identified with a sign that every child is taught to recognize?

The minute one tackles the simplest problem of this sort, whole flocks of issues arise. If a school bus is designed for small children, adults will have difficulty sitting in the seats. Isn't it simpler for restaurants to provide a small chair that fits on their regular chairs, so that a child can see at the table, than to have high chairs? If the parents carry the baby's chair with them, they are almost required to have a car to transport it. Or consider the possibilities and problems of transporting babies — shall we do it with a back-carrier or with a cradleboard so that the baby can be set down anywhere? Or how about swaddling the baby so tightly that it can't roll off a table?

In a slowly changing society, all during one's life the same materials are used, the same tools are relied upon, houses are the same size and shape, and children grow up in the kind of setting in which most, or all, of the details of life have been conveniently worked out and allowed for. Over time, allowances have been made so that clothing and furniture, stairways and doorways, hooks for pans and shelves for crockery, all fit together. If some things are stored up high, provisions have been made for climbing to get them; if babies are laid on the ground at an American Indian gathering, those who attend learn to watch where they put their feet. If candles are used, children grow up to be aware of the dangers of a swaying curtain; and if women wear elaborate coiffures which are not touched for a week at a time, children learn to tussle and lovers learn to make love without disturbing these expensive creations. If food comes unwrapped, as bread does in many countries, there are conventions about how it is to be carried.

But even in very old traditional societies, where the physical settings and the habits of the people have been adapted to each other for centuries, the fit is never complete. Like the disposition, at a cocktail party, of the pit in an olive from a drink that has been fashionable for over fifty years, there are always some social situations which are awkward. At a dinner, some chairs have to be placed near the door; an open door means a draft, while shutting it makes the room stuffy. An open fire in one part of a room means that some people roast while others freeze. But many cultures have found ways to build these discrepancies into the social fabric, and manners are adjusted to such matters as the temperature: the portable electric heater or charcoal brazier can be placed behind the guest of highest rank. Meals in which the supply of food runs out can be made into ways of training small children to pick up social cues fast and not to ask for more.

The very fact that there are so many alternative ways in which living arrangements can be made human, and that most cultures include many of them — although in different proportions — introduces extraordinary difficulties whenever conscious planning is involved. When such planning is done on a large scale, and whole cities are put on the drawing board, the complications become almost insurmountable, and we frequently end up with cities that are mean, cold, inefficient, and unlivable. Or, in another context, we end up with shoes that fit no one, chairs upon which no one can sit with comfort, teapots that don't pour, and people who move without grace. Americans have been characterized as a people who walk as if they owned the world, but part of that confident walk is based on the expectation that, however a gadget works, it can be figured out. On the other hand, English women have been characterized as behaving like duchesses, which suggests putting a postage stamp on without licking it adequately, in the expectation that some kind person will relick it.

The old country men and women who walk with slow, hesitant steps through the airports of the world, speaking no language they can hear spoken, experienced in no detail of the trip, confronted by airline hostesses who demonstrate the use of oxygen masks at lightning speed, represent a combination of inexperience and extraordinary bravery, but a bravery that can hardly be translated into participation in the countries to which they are journeying.

We are beginning — but only beginning — to think about some of the principles which underlie making furniture and dishes and clothing, or building meeting places and transportation systems, to a human scale for people who come from backgrounds very different from those of the designers. At the same time, we must continue to plan for a people who stay at home, in the climate where they have been reared, eating food to which they are accustomed, and sleeping on beds which, if they give them backaches, give them backaches which they expect to have.

But there is almost no place in the whole world today where there are neither people, nor objects people use, nor styles of buildings in which people live, which have not come from somewhere else. The kind of fit between generations of use, generations of craftsmanship, and generations of those who were arbiters of taste and excellence in these traditional things is gone. We need to devise some new way of relating human beings to the physical world around them which will give us, not what we had before, where there were thousands of individually produced variations in life styles, each with its distinctive form of building and living, but something new which can draw on this past diversity and also upon the practices of mass manufacture, which, in so many ways, has made life less burdensome for people all over the world.

At present, the best we know is a little about the interaction between users and the things they use. From the large number of accidents on the steps in front of the new Metropolitan Opera House in New York City, we know that no matter how pleasing their design or how well, aesthetically, they complement the overall design of Lincoln Center, something is definitely wrong with the dimensions of those steps. Attention to the scale of other steps on which accidents did not occur would probably have been enough. Or, mock-up steps might have been built, and a heterogeneous group asked to walk slowly and rapidly up and down them. Professional models might have been photographed climbing and descending the mock-up steps, and their movements might have been analyzed. But each of these precautions takes time, and energy, and money. The more completely it is possible for the architect to draw upon the inexplicit harmonious adjustments of the past, the easier it is. The simplest course would have been to find buildings where approximately the same kind of audience used similarly designed steps and where the fewest accidents have occurred, and then to use the same dimensions. But, here again, as we pursue the matter new complications present themselves: the ethnic composition or the sex ratios of the audience may be changing, the lighting may be different, smog may be increasing, fogging people's eyes, and on and on.

Another alternative is to involve not previous styles, as expressed in buildings, in furniture, and in household items, but the people who will be using the thing under design, a door or a chair or a washbasin, or a new fuel for cooking, or a new kind of disposable pen. If a suitable sample of the group of people who are going to use the new pens is given them to write with, the ways in which the new pen will be used can be observed and fed back into a differentiated manufacturing process. Consider fountain pens: they were expensive, had to be filled from containers which spilled and stained, and leaked at high altitudes; the Soviet airways, in the 1960s, provided me with a neat little plastic bag in which to keep my expected fountain pen. Fountain pens were individual enough to recognize and so less wise to steal; and, usually, one was enough — only rarely did men wear two in their breast pockets. Ball-point pens are cheaper, quite disposable, stain less, require no filling from a bottle — but are impossible to guard against "borrowing" and a type of usage which has absorbed them into the ethic of pencils rather than that of fountain pens. In contrast to the old fountain pens, cheap ball-points are strewn around the affluent world, to be used and discarded.

Watching the users of old objects before the new is designed is a halfway step between relying on the opinions of experts and the opinions of users, whom the experts believe to be unable to choose intelligently because they have never had the knowledge of materials and design that the architect or designer has. The intermediary step that can be introduced is the small, real-life model within which people accustomed to one house style, or one kind of cooking or plumbing, with careful initial instruction, can practice living with the new kind of stove or water closet. If a small group of the future users are brought slowly and selectively into a system of new housing, or new types of schools, taught the technical things which they do not know, given time to learn ways of adjusting to them, a communicable knowledge of how to live with these new things can be developed that future residents can learn.

But if buildings for one group of people are designed by people from another group with a totally different life style, who know nothing, for instance, about different cooking habits, or the need of small children for quickly accessible toilets, the results of such designing will be such things as people throwing their garbage out the window (as they used to do in the open country, where there were chickens to devour it), and small children who pee in the elevators when the nearest available bathroom is thirty floors from their playground.

Deeper perhaps than the need to study the habits of the future users, to adapt their habits to new materials and styles, and to adapt the new materials and styles to them is the question of the degree to which designers respect the future users of the buildings which they are planning. This respect can take many forms: affectionate concern and respect is evidenced when a planner, who cares for his aging parents, insists that the housing designed for elderly people contain the features which, he observes, the parents whom he loves need. Thus, provisions for cooking and bathing which are safe for older people need not be designed by people who are themselves over eighty and likely to slip on a piece of soap in the shower or on an icy outside stairway in the winter.

At present, the best planners are confused by the need to respect the first users of a physical object and the need to anticipate and respect future and different users in the abodes they are designing. But there are solutions. High walls can be designed for the residences of Western Europeans in Moslem countries, which will not distress the Europeans, but which will protect the sensitivities of the Moslems who may live in the same homes later.

Some claim that the only way that an institution can be fitted to the needs of a particular group — students, women, the urban poor, ethnic minorities, guest laborers, or rural people just come to the city — is to let these groups, always seen as disadvantaged, make the plans themselves. And yet neither the gracious Georgian manor houses nor the tall, fair houses of eighteenth-century Europe were designed by the users. The crescents of London and Bath, and the streets of Paris and Bucharest, were built by architects and builders who were, in general, poorer and less prestigious than those for whom they were designing. The style that they created placed a high valuation on such considerations as dignity, lavish entertainment, and privacy. When, however, the same builders constructed homes for the poor, the dimensions were meager and the style mean and cramped, expressing contempt for those of lower status and assumed lack of taste or fastidiousness. When a small house was built for people of consequence, it could have all the grace of a larger house; but the servants' quarters on the upper floors at the back of great houses, or the rows of houses built for serfs, or slaves, or miners, or factory workers, expressed in every line the low esteem in which the future occupants were held.

When building is done for those who are respected, the respect shows in every aspect and every proportion. When clothes are designed for those with lower incomes but equal status, they often have the same lines as more expensive clothes; the saving for leaner pocketbooks should come from the use of less costly material and narrower seams, but not by making the clothes ugly, awkward, and in gross sizes that fit no one. It is interesting that the tendency to design public housing in a gross and inhuman fashion for the unrespected others — the poor, the immigrant, the peasant just come to town — occurs whether the building is in a Free Enterprise or a Socialist country. In Budapest, the great, Soviet-style, high-rise skyscrapers of precast cement stand cold and inhospitable beside some small, friendly high-rise buildings that had been constructed in a style imported from Scandinavia, one which had been worked out by a people for themselves before they had any significant immigration.

This contrast within a single city dramatizes the point that it is not that each group must do every detail of its own designing, when often it has no one with the necessary expertise, but that it is the attitude the designer takes toward the group for whom he or she is designing that is crucial. Thus, men who like women, white people who enjoy American black people, celibates who respect family life, and Caucasians who have taken the trouble to learn how Native Americans want to live can design sympathetically for people they can never be and for ways of life they can never fully experience. But they can do this only if it is accompanied by intimate understanding and affectionate respect.

And, we may well ask, how can someone designing a pot that is to be made by the hundreds of thousands in factories and distributed all over the world ever come to know and love and respect its multiple users? It must be simpler and more abstract than something elaborated for known local uses. And, it is equally important to recognize that the insistence on race or sex or class membership as the best criterion for engineering change is also one-sided and one-sided in a way that obstructs the use of the new for the preservation of the best values in the old. It can result in creating a new educated elite of the "right" race or sex or ethnic background, which is as insensitive to poorer members of its own group as any outsider might be.

Only some orderly resolution of the three modes of designing that exist today — designing at a distance in indifference or contempt, designing by the outsider who respects and knows the people for whom he or she is designing, and designing by one of those for whom the thing designed is to be used — can possibly restore a world of 4,000,000,000 people, in which mass production in many cases is cheaper and wiser, to some semblance of meaningful relationship between themselves and the tools and houses and clothes and utensils that they use. And this resolution must be achieved soon, for the building begins each day.

II

MOURNING, PERU

THE
WAITING
WORLD

BALI

NIGERIA

NIGERIA

BALI

U.S.A.

NIGERIA

INDIA

PIOUS WORK, INDIA

VILLAGE STREET, EGYPT   VILLAGE SCENE, NIGERIA

CAMEL MARKET, NIGERIA

WASHING CLOTH, INDIA

WOMAN, PERU

TWO WOMEN, COLOMBIA

CEREMONY, NIGERIA

GROOMING, BALI

"MASSAGER," INDIA

HOLY WEEK, SPAIN

RELIGIOUS COSTUME, SPAIN

"CARNIVAL," BRAZIL

DANCERS, NIGERIA

DANCER, NIGERIA

RIGHT: WHIRLING DERVISH, EGYPT

HOLY WEEK, SPAIN

STATE FAIR, U.S.A.

PRAYING, JERUSALEM

TELLING BEADS, LONDON

SCHOLARS, TEL AVIV

SAMBURU COUPLE, KENYA   MOSLEMS PRAYING, NIGERIA

GYPSY WOMEN, INDIA

LAUGHING WOMEN, HONG KONG

WOMAN POSING, PORTUGAL

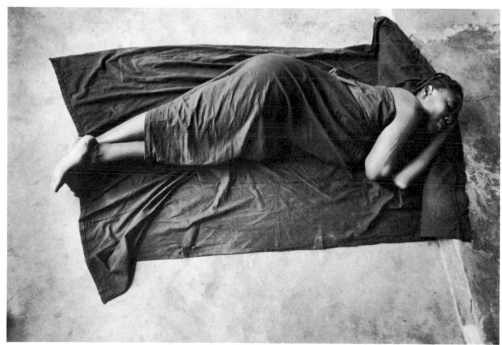

WOMAN RESTING, GHANA

PORTUGAL

IN A PARK, LONDON

DONKEY CART, CAIRO

IN A MARKET, VIENNA

PLANTING SEEDS, BULGARIA

86

POUNDING RICE, BALI

LIFTING WATER, PORTUGAL

POLING SAMPAN, HONG KONG

TOKYO

4

Peoples
in
Need

When World War II ended, two-thirds of the world was relatively untouched by industrialization. Women still brought water from a common well or a distant stream, mules and camels and donkeys pulled simple plows or women worked with digging sticks. Clothing was still traditional: old women muffled themselves with heavy, dark garments; African brides were still wondrously adorned in metal ornaments; in warm countries, people worked without clothing or sheltered themselves from the sun with protective veils. Industrialization touched their lives only at the fringes, in the form of new things to be bought in the market. In mines, in fields, on sugar and rubber plantations, men and women worked at subsistence wages; in other cases, the men went away to work in factories and left the women at home.

Houses were lit by open fires or little stone lamps with languid wicks; people went to bed when dark fell and rose at dawn. There was no electricity. No water was piped into the villages. There were no dentists. The old were given little mortars and pestles to pound their food until it was soft enough to chew with eroded gums. There were no eyeglasses to help waning eyesight, just as there were no strong lights to see by after nightfall. Old endemic diseases flared into epidemics and half the population of a village would die. Babies died like the flies that carried infection to their eyes. Cloth was worn until it fell to pieces. There was no soap, and skin diseases were spread by cherished pieces of scarce cloth. Sometimes as many as 90 percent of the people could neither read nor write. In some countries there were only a handful of graduates from the universities and professional schools of Europe and America.

In the past, people worked very hard, although less hard among hunters and gatherers than among agricultural peoples. People worked very hard in the cities, but most were poor and dependent upon an economy which provided them only with a minimum for survival, only with small, uncomfortable, and unhygienic houses made of mud or stone or wood, or even leaves and reeds — nothing more than huts and hovels in the eyes of the people of the wealthy nations. Water periodically rushed down the center of city streets or culverts that had collected human waste. The efforts of people to stay clean by washing themselves, their clothes, their pots and pans, and their babies in the same water spread disease.

Money was little used, and then only to buy those special things that could not be obtained locally through barter. In many parts of the Middle East, people distrusted banks; in Istanbul the distrust was so great that ubiquitous neon advertising was required to overcome the resistance of those who had recently moved to the city. When people first moved from the traditional countryside, they needed public health services and credit and, eventually, insurance. In time, of course, they would need all the institutions of the modern world, because, as we discovered when we looked closely, all these things go together. The minute wage labor is introduced, other institutions are needed to protect the wage earner from the demands of friends and relatives still living in the older system of barter and kin help. The wage earner needs banks, as does the farmer who wants to use a tractor and improved seeds. Thus, we came to see the world as a vast, unimproved, waiting mass of people who had been deprived of the benefits of our industrialized system, and who wanted all of them.

Everywhere there was what was called superstition: weird rites and beliefs, monks who pored over ancient manuscripts, Yemenites, writing out on their knees from memory, in a beautiful script, old holy words. Superstition was held responsible for the slowness of farmers to change, for resistance to public health measures, to powdered milk, and to bottle feeding. When babies died and mothers sighed, "It is the will of God," public health practitioners bristled with anger. "They seem to want their babies to die," these emissaries of modernity said. No superstitious pretechnological practice had any good in it — although, of course, search teams occasionally found plants useful for pharmaceuticals among the many useless herbs.

In 1953, I went back to study the Manus of Papua New Guinea, whom I had studied in 1928. They were a primitive people, whose exploitive, materialistic culture had mirrored many of our own values, even in its primitive form. They showed enormous enthusiasm for remodeling their own culture, which they quickly perceived as wholly bad, and for transforming it into a simplified version of American culture as they had experienced it through our GIs in World War II. The welcome they gave to literacy, to health clinics, to modern medicine, to the orderly procedures of law (as they replaced threats of violence and destructive, supernatural sanctions) was an enthusiastic reflection of the world to which they had been exposed.

Everything that they had done before, such as using dogs' teeth and shell money instead of a modern currency, now seemed inferior. Indeed, they were shocked to find there was trouble in exchanging their cherished American money for other currencies, for they had gone so far as to assume that the modern world had invented the idea of a single worldwide currency. They rejected a dependence on memory and involvement in endless disputes over facts that they now knew could be recorded in writing. They rejected subjection to the whims of guardian ghosts and instead conducted orderly town meetings by Robert's *Rules of Order*.

The Manus presented a miniature version of the peoples who were, in the minds of Euro-American planners, fully prepared to benefit from a worldwide system of literacy, public health, and democratic procedures. They saw the system which they encountered as wholly good and thoroughly worthy of emulation. They discarded their old ways, their old marriage exchanges, and their burdensome ceremonials. A house building style which had made it possible to build only one or two houses each year was traded for a system of crude mass production which could put up fifty poorly built, but modern, houses in a few months. Everything about the old way of life was tabooed; and the new was de rigueur: trousers were the style, and houses that abutted on the center of the village had to be built of tin. For all this, I saw in the transformation of the Manus at least a suggestion that people could improve their lives by their own efforts as they responded to the challenges of our modern world. The old way of life had been oppressive and exploitive; the old "superstitions" had weighed heavily upon them; young men had remained in thrall to those who paid for their marriages; women had been compelled to hide their faces from their male in-laws; and husbands and wives had been reduced to angry nodes in a net held together by economic exchanges rather than affection or trust.

No country in the world, no tribe, no village, was free, or is free, from many things that some of their members will not want to change as soon as they are exposed to the technical and social inventions of contemporary civilizations. One society will find its youth oppressed, another its women exploited. Houses that once sufficed are criticized because there are no partitions, or because they are too likely to be knocked down by hurricanes or earthquakes. In one society, mothers-in-law are a nuisance, in another there is a bride price, in a third too many ritual exactions leading pious men to devote large portions of their lives to ritual activities. Changes in human behavior have been stimulated throughout the world for hundreds of thousands of years by the excitement that flows from finding out that there are other, more satisfactory, and more rewarding ways of doing something. No way of doing a thing will look wholly satisfactory to its practitioners if they have a chance to compare it with another way that is equally, if not more, efficient, unless, of course, they use that other way to bolster up their faith in their own way by declaring it wholly inferior.

The peoples of the world whom we conceived as simply waiting with great eagerness for every kind of modern improvement did, indeed, reach out their hands for the new methods that could keep babies from dying and give the old the ability to see again. The modern world used this need as a form of self-congratulation. We had sins, it was true, but the sin lay in the fact that we were not sharing our wealth fast enough with the poor, either in our own country or in the new nations. But every reproach against the industrialized world by the unindustrialized served to increase our sense of being on the right track. If artists, who had found traditional ways of life beautiful and thus lamented the loss of traditional dress and dance, learned that these were not ingrained in the people themselves and saw how quickly what had looked like an impeccable taste for color and design vanished as modern plastic bowls with large printed flowers replaced clay pots that had been shaped in beauty over hundreds of years of use, then surely there must be ways of achieving beauty with modern methods of mass production, and surely computers could be programmed to make exact copies of ancient vases when there were no human hands left who knew how to make them.

And moral arguments were advanced against any attempt to preserve peoples' traditional ways of life. Such attempts, it was said, smacked of turning human beings into animals in a zoo, captive primarily for the pleasure of the onlookers. However uneasy we might be to confine the king of the forest to a small, restrictive cage for the "benefit" of urban children who have never seen a lion, there was something even more repellent in asking human beings to live as they once lived for the delectation of artists, social scientists, tourists, political philosophers, and ideologists. Once a people had seen what we had and how marvelous it was, how could they not hunger and thirst after it? And, further, was it not their right? Did not South Sea islanders, never more than a half-mile from the sea, nevertheless have a right to modern plumbing? Was it not ethical that when we found a New Guinea people afflicted by a terrible disease, kuru, the laughing death, believed to be inherited, we should try to quarantine them in the wilds of New Guinea, or was this a form of racism? Was not the attempt of the United States to keep out European refugees who showed tuberculosis scars on their lungs inhuman, a wicked form of isolationism? So it was judged from Sweden, which proclaimed itself ready to take — and cure — refugees actively suffering from the disease.

All of these arguments culminated in the insistence that all people, of all races, of all cultures, and of all circumstances have — or should have — the same opportunities, rights, and privileges, and that the new nations should have positions in the world equal to those of the older nations. It was claimed that they would have achieved salvation by industrialization long ago if they hadn't been held back by the colonial regimes. Poverty — often newly discovered among people who had lived with too much pride in their own self-sufficiency to feel poor — was someone else's fault, and wealth, as it was defined in the postwar world, was to be sought and accorded respect by the poor as well as the rich of the world.

In spite of their eagerness for the new, the quicker, the more efficient, the less laborious ways of doing things, the peasant and primitive peoples of the world did not change as quickly as those whom they had called in to help them wished. There was much conservatism, much resistance. In the frenzy of modernization, this was seen as an evil. In almost every traditional society, there were schisms between those who embraced the new ways of life and those who clung to the old. In Turkey, where Atatürk had been able to introduce enormous changes by fiat, sweeping away the old script and unveiling the women, conservative rebellions developed in the villages. In Indonesia, where the religion of Islam had been tempered by the greater tolerance of earlier religious beliefs, fanatical sects of Moslems who opposed everything that was foreign sprang up.

In the more industrialized countries, there were cults of all sorts, groups of people who wanted to keep or reinstate earlier ways of life. The foreign experts deplored the conservatism of the people who refused the new seed or the new breeds of cattle. The educated leadership hung their heads in shame over the backwardness of their people. When change is in the air, conservatism appears as a great evil to those who are riding the crest of change.

And it is true that if people are going to make a drastic change in their lives, send their children to school all day, work for wages, introduce modern methods for the treatment of disease, wear manufactured cloth and use modern tools, then it is easier to change all of these things at once, rather than in a slow, piecemeal fashion. The immigrants who come empty-handed and prepared to learn bring less confusion than those who try to bring their farm animals into the city. People who accept schools at all get more from them if they stop lamenting that there are no children to help herd the animals or care for the babies anymore. All the new things — new forms of fuel, new stoves, and new cooking utensils — go together; garments made of cloth must be washed with soap and mended with a needle and thread. Constructive and effective behavioral change goes better when whole systems of interrelated behavior can change at one time.

But we have learned today that not all efforts to resist change are wrong-headed. For many years in America, we have heard of the terrible conservatism of physicians who refused to consider new remedies or new theories of disease; but, on looking back, the excessive willingness of physicians to accept wonder drugs, mechanical methods of diagnosis, systematized and depersonalized hospitals, bottle feeding, hospital delivery, and extreme measures to add an extra day or two — or hour or two — to lives that are over may in the end be found to have done even more harm. Similarly, teachers have been berated for their failure to be innovative, for their rigidity in clinging to old methods of teaching. But the exaggerated eagerness of educators in America to seize on each new idea, however untried and unthought-out, may have been even a greater hazard.

In times of affluence associated with change, resistance to change goes unrewarded, or even punished. But in times of stress, people turn back to the older, tried methods. So on small South Sea islands during World War II, when people ran out of woven cloth, those who had not forgotten how to make bark-cloth were able to make it again; and, deprived of electric light, they could light their houses, however dimly, with candlenuts.

But throughout the world, the new was driving out the old, and the city, which stood for the future, was a lodestone, drawing people away from the countryside long before there were jobs for them. As new methods of farming spread, there was less need for farm workers; as new methods of controlling old endemic diseases spread, more children lived, and, suddenly, there were more mouths in the country to feed than anyone had planned for. The city offered an opportunity to escape the exactions of rural life, the daily round of caring for animals, the anxieties and disappointments that go with dependence upon the weather and world price fluctuations that seemed as arbitrary as the weather, the repetitive admonitions of relatives and neighbors who were certain of one's expectations and limits, and the lack of privacy in living among those who were alert to the slightest change in one's behavior.

Cities have been a matter of pride ever since it became possible to build them. Here large bodies of workers could be gathered together, and great monuments could be built, great palaces and temples executed, art and thought could flourish, and large numbers of gifted people could meet and exchange ideas. In the cities, the unexpected might happen any day, new avenues of advancement and changes of status that were daydreams in the country could be realized. In the cities people become accustomed to being fed by others, to buying clothes, to living in houses that were designed and built for someone else by someone else. In good times, the pull of the city has been irresistible. In bad times, those who could returned to the land, where a little something could be planted or gathered. In wartime or times of national disaster, the soup kitchen does not seem so terrible to city-folk as the famines on the land from which people have always drawn their livelihood seem to country people. As soil was eroded, as land was overgrazed, peoples all over the world were driven toward the cities. And the cities, in turn, needed people to work in the new industries and to buy what was made there. Modernization and urbanization go hand in hand; the acceptance of one has required the acceptance of the other. In cities, there is always the hope that one may escape from working with one's hands; in the country, the farmers have been tied to their daily toil as inescapably as their wives have been tied to childbearing.

When the plow and the cultivation of grain made it possible to store food in sufficient quantities so that a small number of men could be freed for other pursuits, civilization began. Cities appeared first, perhaps, in locations that were set aside for festivals or around the residences of kings, to which great numbers of people came for short periods to bring offerings and tribute and worship, much as devout Moslems today still make pilgrimages to Mecca. Later, more developed urban organization made it possible to establish granaries in the cities, to develop craft specializations, to engage in larger-scale trade, to carry on military adventures, and to expand empires.

Each advance in social organization meant a more extensive division of labor; rare jewels and precious metals could be brought from a greater distance, greater social stratification occurred, greater distinctions developed between ruler and ruled, free and slave, and rich and poor. And, indeed, down to the present, there have never been cities where there were not some unfortunates, some castaways from society, dependent upon the alms of the religious, upon pittances scavenged from the garbage heaps of the better fed, or cities in which some of the children did not grow up to be beggars, wastrels, thieves, or thugs.

The ability of human beings to manage cities has waxed and waned; sometimes the sewerage systems have been better than others; the magnificent system of sluices in the city of Tehran, through which the water rushes diurnally, gives a rhythm to the city; however, it is no longer appropriate for the kind of modern urban life that is lived there now. Roman cities of antiquity were cleaner than the European cities of the Middle Ages, when chamber pots were emptied from upper stories upon the heads of passersby, but the early Roman senate sat in unwashed wool. In Europe and America, in the nineteenth century, epidemics from waterborne diseases were common, but city fathers denied them to protect their investments. Cities have been the glory — and the defeat — of one civilization after another; periodically, they became top-heavy and the home of too many who did not earn a living, exhausting the supporting countryside, rising and falling with the fate of the dynasties of which they were the center.

Although cities were the pinnacles of civilization, the abodes of kings, the sites of the most developed and elaborate worship, the seats of learning, and the centers of innovation, historically, their populations were small in proportion to the number of people who lived in the country and provided food for them. Cities could rise and, when they fell, the populace could disperse into the countryside again; the effects of plague and famine could be absorbed by the peple who lived on the land.

Cities fostered distinctions between the sophisticated and traveled and their country cousins who remained on the farm. Even in the very simplest societies, we find that similar distinctions are made: thus, differences between the traveled and the untraveled, the more skilled and the less skilled, and the migrant and the established residents are clearly perceived. In traditional societies, a kind of balance obtains between the admiration for the more traveled and the pride of those who stay in one place and become thoroughly identified with it. In some parts of the world, descendants of families who have lived in the same place for several hundred years speak of themselves as immigrants whose real home is a thousand miles away, thus combining the pride of belonging where they are with the pride of having come from somewhere else. This is the typical ideology of the colonial; the successful descendants of colonists, and often those who have long since ceased to be successful, view themselves as the oldest comers from a still older land. But their situation is balanced by that of the later immigrant, who comes to a new country as a poor man, dependent upon slightly richer relatives, and although forced to take the lowliest, worst-paid jobs, nevertheless can return to the poverty from whence he came, richer, more traveled, to be treated with new respect. In the ebb and flow of populations, those who flee from the hunger of a depressed and neglected countryside, who are forced out by increased mechanization or increased population or both, who are dispersed by war and revolution, who are refugees from one political system drawn to a more attractive one, can be contrasted with those who elect to move, to join another group, and to seek a more promising way of life.

Within the cities of the world today — especially in the new nations, where the emphasis has been entirely on the cities and where the countryside has been grievously neglected — we find both groups: those who have fled in despair and often have remained despairing, and those who have traveled in search of a better, more interesting life. Sometimes the two groups come from the same people; but the most energetic are often the ones who, rather than fleeing in despair, seek new conditions and characterize their departure as enterprising and adventurous. As the world has become more interconnected and as ships and planes could carry people farther away, the two groups, the despairing fugitives and the adventurous settlers, have gone farther and farther from home. These movements have been so important that it is interesting to speculate what the world would have been like if ships had not been invented and all travel had been on land, on foot, or at the pace of the horseborne hordes of Asia, who moved very slowly, stopping for some time among the peoples they conquered.

Some of the most vexing problems of the modern world have come from the clash between peoples of very different physiques and living habits, who find themselves, as a result of immigration, living next door to each other. Thus, today, we find old and new descendants of African, European, and Asian origins, all living side by side, especially in the lands that they discovered and settled last. The strangers and newcomers who come from close-by do not make such a demand for acceptance as those who contrast in skin color, in hair form, in styles of body movement, in preferences for strange foods (that give off strange smells when cooking, offending the noses of their neighbors), and in different styles of quarreling and reveling that fill the air with bizarre and unendurable sounds.

Cities have struggled with their immigrants, the ones who seek work and the ones who flee from hunger, as well as those who come to stay briefly. The city has struggled with those who are out of tune with it in one way or another and has found that these forms of struggle recur through history. The walled city locked its gates at night and forbade peasants to stay overnight. Ancient Peking had sections separated by walls with gates that could be shut at night. Today, in Manila, there is a separate city, Makato, for the fortunate and privileged, with its own water supply, its own electric system, its own supermarket, and gates that are guarded. Periodically, cities become unmanageable; there are too many people with no work and no source of income. Then the wealthy take measures in an attempt to reestablish control. As once the people of Sardinia tilled their fields by day but slept huddled together in towers at night, so in later times, people within cities erected various kinds of refuge, guarded and barred against the desperate and marauding segments of the population.

Today, all of these conditions of ancient times — people fleeing plague and famine and war, people seeking adventure, people moving toward opportunity — are exaggerated, both by the tremendous increase in population and by the ease of travel. As countries become industrialized and raise their demands for new workers, at the same time that they educate their children to a point where they no longer want to be manual workers of any sort, places open up for migrants from the less-industrialized countries: thus have Greeks and Italians gone to Switzerland and Germany, Yugoslavs to Sweden, Algerians to Italy, Indians to South Africa, Puerto Ricans to New York, Mexicans to California, and Ukrainians and West Indians to Canada.

At the same time, where a disproportionate number of the educated have been trained for nonexistent jobs, they become surplus in their own countries and then emigrants, entering the higher echelons of the social systems of other countries. Thus, Indian, Pakistani, and Latin American physicians staff the hospitals of London and New York; Gujaratis follow where poor Indians have migrated to manage their trade; highly educated Japanese, Chinese, and Koreans remain in the United States to take positions of eminence in university life. Sometimes the migration of the very poor and uneducated and of the highly educated and sophisticated may take place simultaneously, and people who have looked down on the new immigrants may find themselves waited on in a restaurant by members of one class and operated on in a hospital by members of the other class, both of the same ethnic group, the first dependent, timid, and propitiatory, the second often arrogant and overweeningly proud.

The implications of these contrasts in everyday life were vividly illustrated by the case of a small American boy, the dark-skinned child of a blond father and a brunette mother, the eldest sibling of a family of three, who had two handsome, blond younger brothers. In the public school which he attended, he absorbed the prevailing attitudes of the white children toward black children and reacted with fury when his mother became interested in civil rights and took up the cause of the black children. He became so unmanageable that his parents transferred him to a private school, only to be confronted on the first day of classes with the fact that his teacher was to be a tall, black, West African woman. But with all the pride West Africans carry with them, stately and assured, she took the angry little boy by the hand, and he, who had blamed his rejection on his sallow dark skin, felt fully accepted for the first time in his school life.

But the opposite may happen. In New Zealand today, there are many Polynesian immigrants from the Pacific islands who come there for a higher education. New Zealand's own original population, the Maori, is also Polynesian; they have become proletarian workers in the cities where the universities are located. The newcomers, specially chosen, proud because of their rank and their superior performance in school, form a group that appears superior to the "indigenous" Maori population of New Zealand, which contains the whole range of skill and intelligence found in any community. As a result, the estimation in which the indigenous Maori as a group is held is lowered because the specially selected newcomers are, on the average, so much more successful.

In America, darker Mexicans and Puerto Ricans cling to their Spanish tongue for fear they will be classified as the discriminated-against American Blacks, while lighter-skinned Spanish-speaking immigrants hasten to take on the manners, appearances, and language of white Americans. This has led to the development of strange categories like "not not-white."

Cities that have large populations of conspicuously different racial or ethnic origin build ghettos which wall in or wall out those who seem too different from the rest. Sometimes the less accepted are progressively banished from the city to the outskirts. So in South Africa, Blacks are banished to the suburbs with poor facilities while in America it is the wealthy, middle-aged, and privileged population that has chosen to move to the suburbs, leaving the cities to the poor and disadvantaged, to the young, and to the very rich.

In the cities of the newly developing countries, as primitive and rural people have crowded in, there has usually been no housing to receive them. Where Sweden houses her Yugoslav foreign workers in the less-successful, experimental high-rise apartments, Caracas, building one high-rise after another, has failed to keep up with the hordes of newcomers, who build their own little shanties on the hillsides all around the city and in gulches within the highly sophisticated city itself. At night thousands of tiny lights twinkle from these hillsides, contrasting vividly with the lights of the central city.

In the new countries, generally, there have been attempts to deal with these masses of newcomers, first to give them roads, water, and electric lights, and then to let them gradually substitute better housing as they prosper. But this has had the effect of turning all of those who go to the city for work into permanent residents, a mass to be fed and sent to school and given some medical care. This tends to make them into people who will never return to their villages, to which bad times in the city might send them back. And so the industrializing world is trapped on a spiraling escalator. As cities raise the standards of living for their inhabitants — better schools, better hospitals, better provision against starvation — more people flock into them, in turn strangling the city's power to provide for those within its gates. Then, the cities turn against the newcomers, once welcomed because they provided cheap labor; immigration barriers are raised and all sorts of provisions are made to screen immigrants. The immigration — legal or illegal — of desperate, hungry Mexican workers is opposed by Mexicans who have become U.S. citizens and trade union members; the foreign "guest" laborers are sent home from the cities of Europe, back to the poverty that drove them out in the first place.

After twenty-five years of emphasizing industrialization in the less-developed nations, and automation and labor saving in the old, industrialized countries, and after twenty-five years of trying to spread the methods of the commercialized agriculture of the grain-exporting countries — the United States, Canada, and Australia — and of the collective farms of the Socialist countries, the world is facing a terrible reevaluation of the policies that seemed so promising just a quarter of a century ago. For those who still believe that Euro-American–style technology has the answer, the contrasts become ever sharper. Thus, the head of the World Bank, in an hour-long broadcast laying out what raised productivity could do to ameliorate the spreading poverty and despair in the world that has accompanied the population explosion, illustrated his cheerful hopes by proposing what could be done for an area in Africa where miserable people go blind because of insects that attack their eyes. So as faith in technological solutions grows weaker, paradoxically, more dramatically terrible conditions have to be invoked. The broadcast did not mention the fact, which has been well documented by a blind anthropologist who could venture into a similarly infected region in Mexico, that the people who live in such areas, knowing that blindness is an almost certain possibility in the future, are afraid to move away. If they stay where they are, they will at least know the trails to their gardens, and there will be children to help them. Similarly, miners continue to live on in Appalachia after their mines are closed down, unemployed, neither prepared nor willing to face any other way of life than the dangerous and exacting one that they are used to.

Thus we see that paralleling the growth of the shantytowns that surround the new cities there are the peoples all over the world who cling to their former way of life. Some cling to herding, which has become much more difficult because newly established borders prevent them from taking their herds to the most suitable pastures. Some cling to their hunting grounds, although newly enunciated rules prevent them from hunting. Some stay on cultivating little plots of land, although the possibility of work in distant mines or plantations now lures their young men away, leaving the women and children and the very old to till the impoverished fields. For some the introduction of compulsory school is turning half of the young people who do stay at home into despised dropouts, looked down upon by their successful siblings and neighbors who have gone away to school, and who, in turn, despise the rural life which they are forced to live. We saw all these peoples as waiting for a new way of life; but today many of them know only despair.

5

Widening
Categories

By the mid-1960s it was clear that the simplistic dream in which the industrial countries — among whom the Soviet Union was now included along with the United States and the nations of Western Europe — would be able to spread their benefits over the entire world was not turning out as expected. The conference on Church and Society, held by the World Council of Churches in 1966, dramatized this stage in the postwar struggle for greater justice and equality in the world by what was, essentially, a Western constituency that operated mainly from ethical, rather than solely economic, premises. I say "Western" here with emphasis, because this group — like most of those they represented — had yet really to include the whole world in their thinking. Japan was not yet seen as part of the worldwide industrial complex; the People's Republic of China, about which very little was known, was referred to as a "developing country," neatly pigeonholed with the new African states, just as the Church before Vatican II (1962–1965) had pigeonholed the members of the high traditional civilizations of Asia with the "pagans" of the jungles and of the remote Arctic wastes.

That gathering was the first World Council of Churches conference at which there was more than token representation from the developing world. Nonetheless, the discussions and arguments there were conducted in the old terms which both Christianity and Socialism (viewed by many as a secular form of Christianity) had set up. The rich owed help to the poor, the well owed help to the sick, and the wise owed help to the ignorant. This was the old Judaic formula in which, however, "the chosen people" were taken to be that people who accepted the "right" faith and, as a consequence of which, were enjoined to practice it toward the rest of the world. But, at the same conference, the representatives of the young, developing nations insisted that any attempt to curb environmental pollution, which was beginning to worry the West, was only a guise under which the West would, for instance, cut down on helping them pipe water into their villages; the philosophers of despair railed against what modernization had done to Europe; the older-style pietistic religionists pleaded for closeness to God; and the exponents of what technology could do when used for the right ends fought against those who said the fruit of modern technology was the death of the soul.

All around the world there were stirrings of doubt and disillusionment. The system of treating each nation-state, large or small, rich or poor, as a unit within which progress could be established was boomeranging. Warning voices were heard on every side identifying the problems, but they were not yet seen as symptoms of a common world condition:

The gap between the poor and the rich countries widened.

In the developing countries, the contrast between the few rich and the many poor sharpened.

In the rich industrialized countries, the creation of large, affluent, middle-class populations was using up an unconscionable share of the resources of the world.

If the poor countries charged a fair price for their raw resources, the rich industrialized countries would simply retaliate by making synthetics.

The cities of the world were becoming unmanageable.

The countryside was pouring its starving millions into the cities.

The hopelessly parochial standards of building construction, originally devised for Great Britain and Europe, were inhibiting the design of buildings appropriate for countries with different climates and different life styles.

The spread of nuclear generators was endangering humanity, which had no idea of how to dispose of the poisonous wastes that were accumulating.

The kind of education that was spreading across the world was not producing more appropriately educated populations but rather was leading to hundreds of millions being branded as failures.

The world was increasingly controlled by multinational corporations that manipulated governments with little restraint.

South America represented a hopeless dilemma in which peaceful progress was unlikely, and Fascism more likely than Communism.

As long as there was one colony left in the world, colonialism had not ended.

Some argued that the only solution to historical injustices was for the nations which had been depleting the vast resources of the undeveloped nations to make restitution to those nations so that they might achieve their own advanced state of materialistic and technological satisfaction.

Others argued that the only solution was for the Western Capitalist world to give up its labor exploitation and profit-mania and become Socialists. This position was continued into the 1970s by revolutionaries in the Free Enterprise world.

And yet, at the same time, a kind of reactive interest in religion was appearing in the Socialist countries of Eastern Europe.

Each group of complainants was calling attention to a part of the world system, and each had a different remedy. There was hardly any recognition yet that our intercommunicating world was, in fact, *already a system,* in the sense that any change anywhere might affect the whole. But even the horror scenarios of nuclear or biological disaster, which were produced from time to time, failed to shake the old habits of thinking about parts of the world as autonomous and self-determining, and others as victims, customers, sources of raw materials, or ideological converts.

If the United Nations had little police power — and it had only a little and no sanctions that could prevail against the special interests of the large countries, the large corporations, and self-interested local leaders — then, it was argued, we did not have one world, but a world still polarized between the two versions of the Western ethic, with China an unknown whose presence was felt but which had, as yet, to be included in our thinking about the world. The old political and economic ideas still dominated the scene, as there was no worldwide government and no worldwide agreed-upon economic system. It was only possible to think of pieces of the world, of aspects of the behavior of nation-states or of widespread economic enterprises, of the kind of barter preferred by the Socialist countries, and of the currency deals of their rivals. But the dangers from stockpiled weapons, from the exhaustion of irreplaceable resources, from pollution of the upper atmosphere, from the exhaustion even of ordinarily replaceable resources — whales and anchovies and cod — were all mounting steadily.

The next step was to see the whole world as a single fragile system — as it had appeared to be when glimpsed from the moon — where the bell tolled for each human being when it tolled for any one of them. The old vision of the poets and philosophers of one world was still accessible only to a very few, many of whom were very young. And yet the old state of humankind, composed of small groups, of larger groups, of very large groups, all protecting their boundaries, winning and losing in contests of two. sides with many players, still obsessed us.

Meanwhile, the primitive and traditional behavior of the Third World's peoples began to look very different to a world where all forward-looking people were worried, whether they were multinational corporations concerned about cutoffs of raw materials, the expropriation or nationalization of plants and mines and the blockage of trade routes by local wars, or unstable governments that might rise or fall because of some event over which they have no control, or concerned human beings trying to grasp the extent of the human misery around them.

This process of worry picked up the concern begun in World War II when the human spirit was devastated by confronting the incomprehensible and unbelievable enormity of the mass slaughter of millions, the elimination of whole cities by fire-bombing, and the concluding horror of Hiroshima. Skyscrapers that were the background to growing shantytowns, which bred crime and increased human misery, somehow did not look as majestic as they had in the 1960s when they were seen as ennobling the new African cities. A poor family huddled together on a weather-beaten veranda no longer seemed to awaken us to the promise of rural electrification, as once it did in the days of the New Deal, the Fair Deal, and the Great Society. There was less faith in the welfare state, a successful European community of nations, the clearing of the Siberian lands to solve the hunger problems of Eastern Europe, and the miracle rice which would, all by itself, stem the expansion of China in Southeast Asia.

It is astonishing how different misery looked when we were certain that it was going to end, that poverty would be abolished, that the despairing, with no place to sleep except the ground, would soon have houses that would be warm and adequate, that the little, shrunken bodies of starving children would soon be full of warm and nourishing soup. The disillusion mounted while the United Nations and its agencies were fostering a growing world ethic which affected even those nations who flouted it, while each agency pursued its separate hopes, for literacy, for scientific and cultural development, for economic development, for the International Biological Program, for the peaceful use of atomic energy, for the eradication of disease, for combating hunger and replacing it by plenty, and for protecting the endangered environment. Efforts were made, even if the problems seemed more and more discouraging.

But the comforting belief in low-cost panaceas prevailed. It was believed that pollution could be dealt with easily by a few legislative devices, just as deaths on the highways could be greatly reduced by obligatory seat belts. Insurance systems could be devised to protect us from uncomfortable or unbearable catastrophes. In the United States, as medical malpractice judgments rose, costs to insurance companies, and to doctors and patients, mounted, it was hoped that federal insurance would solve the problem. Nuclear power plant accidents occurred and raised fears of enormous damage; but government insurance would deal with this too! The successful — although largely symbolic — replanting of one almost destroyed marsh gave hope that the process of reconstituting a ruined land need not be so impossible after all and that perhaps Lake Erie wasn't so much dead as just miserably sick.

The pictures in Part III show a view of the city as a trap which is destroying human life, rather than as the pinnacle of human hopes. A woman lying on a piece of cloth on the ground presages a time when a billion more people will be homeless. The crazed look on the faces of the men behind barbed wire is not a picture taken in a concentration camp, but simply one of men excluded from watching a game of soccer. The drowned man, lying where he was pulled from the water, with candles by his head, suggests a world in which there will be no space to bury the dead, or where famine will dictate burying thousands of bodies in a common grave.

The pictures of the hungry and miserable call attention to a world in which the *New York Times* could publish an article on triage, the extension to living and well people — who happen to be caught in a world where there is a famine — of the terrible decisions which an inadequate medical corps must make with regard to the wounded in wartime. The lifeboat metaphor keeps cropping up, a not unexpected sequel to the faulty image of the earth as a space*ship*: i.e., a nonliving, completely man-made entity. It is as if, having accepted the notion of our world as entirely artificial and man-made — and thus very fragile and sensitive, and inadequate for the task of survival — now our only recourse is to throw some of the passengers overboard when the "machinery" goes awry.

It took the energy crisis and the hunger crisis of 1973–1975 to bring home to the world how interconnected, total, and vulnerable is the system within which we live. The whole world, everywhere, was affected by the changes in the price of oil and the exhaustion of the reserves of food. Producers and consumers alike, schoolchildren who had to travel long distances to school, factory owners and landlords, poor farmers in India and rich farmers in Nebraska, all felt, simultaneously, the worldwide repercussions of a failure to think of the world as a whole.

There are many reasons why the energy crisis came as such a shock, even to the prophets of doom. The One World concept had its roots in the great religions, which contained such an element of universality that their missionaries could cross any border and their rituals could be practiced in any country. True, in the modern world, some of the inconvenient practices had to be eliminated; it is hard for an orthodox Moslem to pray five times a day in a Detroit factory and for workers who travel four hours to and from work five days a week to get up for an early mass on Sunday. There are few places in the Christian world where young men can, like Buddhist monks, retire for a time and then emerge again into everyday life. Catholics have given up fasting from meat on Friday, and mass can now be attended in the evening without having fasted since dawn. Women are no longer required to cover their heads in church, priests wear secular dress, nuns wear miniskirts, and Moslem women go unveiled. Most nations have been secularized or, at least, have attained a second level of ritualization, which some people take for secularization, like burial in the Great Wall in Moscow.

But in spite of the reduction of ritual observances, the basic universal ideas that made the great religions survive still dominate human thinking: the idea of the fall of man from an original state of purity or innocence; the possibility of attaining, at some time within the realizable future, or in some future state, a condition of peace and harmony; and the notion of apocalyptic catastrophe brought on by the wickedness and intransigence of human stupidity, avarice, and the propensity to dominate, exploit, and destroy other groups of human beings.

It is because of the appeal of these large, sweeping views of sin and salvation that we call some of the religions of the world *great* religions. They contain the idea of the Brotherhood of Man and of the overarching protection and sovereignty of Divinity, within which race and class, sex and nation, and wealth and mother tongue can all be treated as less binding than the shared faith and the hope of heaven. Periodically, little religions appear, expressing ancient religious attitudes, which tie the peasants of isolated communities to the service of some local saint, transformed thereby into the presiding deity of a place; and there is the tendency to create new local cults and establish new local shrines. Most of these local cults have many of the same qualities as the great religious systems within which they appear, especially the idea of conversion. Although Judaism, the religion of the chosen people, requires that its religious leaders warn away rather than proselytize, and although Hinduism will incorporate a group provided caste rules of mutual refusal to marry or mix too closely together are observed, even they — in our strange, intercommunicating modern world — periodically spawn missionary movements.

Ideas of the immutability of ethnicity, that Gallic wit, African rhythms, Irish humor, and Chinese inscrutability are inherited the way eye color and hair type are inherited and do not depend on learning, are widespread; nonetheless, these qualities are now attributed to larger and larger groups. The kind of religious system in which the gods of one lineage differ from the gods of another lineage, or in which each small, localized clan is under the tutelage of a local two-headed rainbow serpent, which cannot be exported as a belief nor shared at a welcoming shrine, are outside of most people's experience. Fellow Christians, fellow Moslems, Communist comrades, the sisterhood of all women, the young people of the world (who establish rapid communication through shared songs when they meet), the Third World (a term which lumps together everyone who is not of wholly European origin) — these huge categories, none of which contain fewer than many millions and some of which contain half the human race, are replacing the differences that once characterized small, isolated peoples who had not yet invented or borrowed systems of trade, shared rule, or shared law.

Imperceptibly, these conceptions of our common fate as children of God, brothers in a revolutionary struggle, sisters in oppression, youth in rebellion, siblings through the lack of a white skin, have been taking hold of the imagination of the modern world, transcending borders just as once the great religions of Asia spread over the world. All workers, all miners, all physicists, all chess players, wherever they were born, whatever language they speak, are gathered into universal categories for the statistics of encyclopedias, for propaganda purposes, for international bodies who claim common cause with millions and who urge their members to go on pilgrimages or to world conferences together.

Typically, these large categories can be broken down into other categories: there are white, black, and Asian Christians, Soviet and American chess players. Games that have become universal and international become renationalized in the Olympics, and then South Africa and New Zealand ban each other's athletes because the other country has included an attitude toward race in its idea of athletics. But all of these are shifting categories; South Africa may repent and permit black teams to play white teams, while preserving apartheid within teams; guerrilla fighters and terrorists can be reinstated as ordinary patriots after a few years of respectability.

The intelligence agencies and secret police of regimes that rule their people by espionage continue to look for small but certain signs of subversion or for membership in some unacceptable political cult, but the openness of the world today is illustrated when we consider that there are few places where the ownership of a copy of the *New Republic* or the *Nation,* a rosary, or a Bible in the vernacular would be seen as signs of subversion. The fierceness with which political police may react to some current political fetish gives young people growing up at that moment the sense that these prejudices have existed for eternity. But, by and large, the worldwide communications explosion has done much to reduce the authority of those who would attempt to restrict a people to a single official view.

Meanwhile, the ease of travel, the rapidity of communication, the exchange of visual images that are intelligible without words, all provide the context within which shared interests, shared grievances, and shared hopes of heaven grow. As planes from any country in Europe reach a midpoint in the Atlantic, their takeoff air control center ceases to follow their flights, and the air control center of their landing on this side of the ocean takes over. Without this quiet cooperation international flying would be impossible. We accept the arrival of a plane at a foreign airport just as we accept the arrival of a letter with a foreign stamp. Only when these international systems are disrupted, when a plane is hijacked, or one goes down because of the misinterpretation of instructions to a pilot, or letters have to be smuggled into some country that has been put beyond the pale, or our passports suddenly contain a new sentence, "Not valid for travel in Afghanistan," do these myriad international arrangements even reach consciousness.

We are constantly surprised when we encounter blocks to wide communication. It surprises us enormously when we hear that the Soviet Army did not have the equivalent of V-mail during World War II, and that Russian women sometimes did not know for years whether their husbands and brothers were alive, or that Chinese soldiers fighting in the Korean War could not find their way back to their homes because many of the remote villages had no nationally recognized name.

Now we expect travel agents to know which diseases are endemic or specially feared in the different parts of the world, so that if there is a possibility that one may want to stop off in country x or country y, cholera or yellow fever shots can be taken. The postmistress of a tiny branch post office, where tea and cigarettes are sold, is expected to know the right postage for mail to any place in the world.

The farther from home, the larger the category with which our minds must work: Yorubas and Ibos become Nigerians, and, as they move from Lagos to New York or Moscow, Africans. Californians become successively Americans, Westerners, and Whites as they move about the globe. This again produces the experience of widening classifications of membership in new and unexpected categories, and greater fellowship with many more people.

Jews from North or South America are called Anglo-Saxons in Israel; Canadians often pass for Americans in Europe if no Americans are present; Middle Easterners in America accept the blanket term of Syrians; Sicilians, of Italians; American-born Franco-Americans are Americans in Canada and French-Canadians in Maine.

After some major event such as a war or revolution, when new groups of enemies and allies have been formed, older identifications will often re-appear, as happened in the United States during World War II, when Italians were as vociferous in wanting to submerge their Italian origins as they have recently become in wanting to proclaim them, for political purposes, in New York City.

Thus, Puerto Ricans refused separate treatment because of their language when they were first recruited in World War II, but when they were put with English-speaking Americans, they then insisted with equal firmness on being treated as Spanish-speaking Puerto Ricans.

The mismanagement of an international problem, such as the energy crisis in 1974, produces rivalries and enmities where there had been none before; nations with oil and nations that depend on imported oil, states in the United States that have gained inhabitants, making a state oil allocation too small, and states that have lost population and thus are getting more oil than they need become competitors. The feelings of alliance or enmity that are generated may become strong enough to provide a political base for some new shipment of arms around the world or for some current election in the United States.

The significance of the way these shifting categories are making us members of One World can be seen in the new habit of defining both sides of any dichotomy with equal specificity. Thus, the old "Christians and Infidels" has become "Christians and Moslems" or "Christians and Buddhists"; "Americans and foreigners" has become "Americans and Italians" or "Americans and Russians" or "Americans and Japanese." Perhaps some of the changes of categories which define moral positions are even more striking; instead of a pejorative term like "homosexual," with the suggestion that the sexual orientation of the rest of the population was simply normal, we now have "gay" and "straight." Where we once distinguished only "vegetarians," we now have "vegetarians" and "meat eaters." In the economic field, instead of terms like "exporting" and "importing" countries, we now have the "oil-producing" and the "oil-importing" countries or the "oil-rich" and the "oil-poor." In other cases, graded categories have developed: the "over-developed," the "developed," the "less-developed," the "developing," the "under-developed." There are distinctions made between "big," "middle," and "small" powers. In each of these an entire universe is contemplated and divided up.

Sometimes these attempts at large-scale reconceptualization may take the steam out of groups who once lived on competition or controversy. This has happened among Christian denominations in New Guinea, where opposing missionaries used to denounce one another's doctrines as leading straight to the flames of Hell; now, under the aegis of ecumenism, they sit down together and find that, somehow, their whole reason for being there at all has vanished. Similarly, large corporations, which, while they are growing, are competitively stimulated by the success of their rivals, become so big that their interests shift from successful growth to matters of power and prestige, and their productive capacities and dividends suffer. Thus, we may, I think, expect that shifts in alignments, the breaking up of large units, and their reuniting under other conditions will be a continuing condition in the world. The main difference from the past will be that in some way the whole world will be included in the way in which the members of any group think about the world and in the way in which the members of any group are thought about.

III

URBAN DEVELOPMENT, CARACAS

THE FAILURE
OF THE DREAM

PRAYING, KYOTO

SLEEPING, CALCUTTA

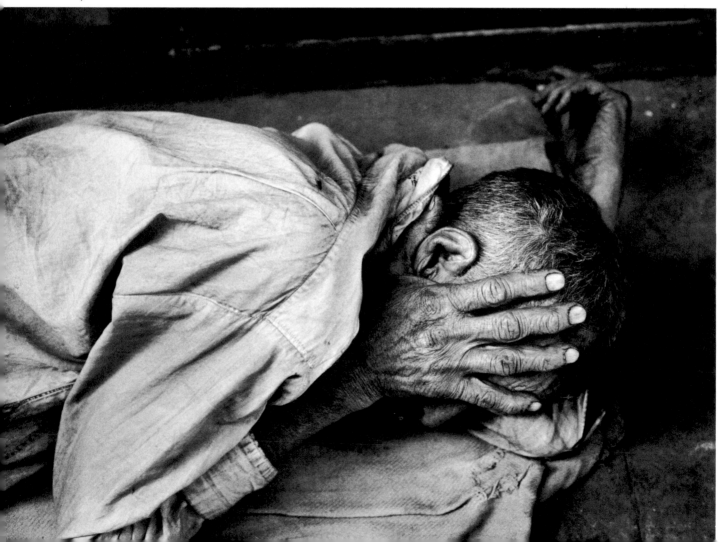

WAITING, PAKISTAN

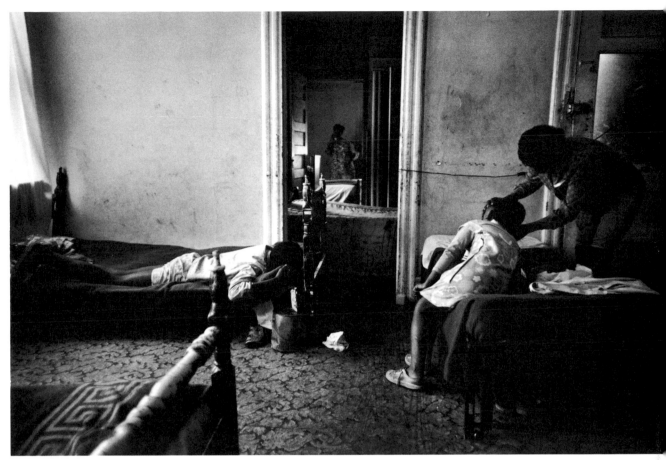

RESTING, U.S.A.

POSING, EL SALVADOR

WAITING, BRAZIL

WAITING IN LINE, MOSCOW

WAITING, NEW YORK CITY

U.S.A.

U.S.A.

U.S.A.

A BEGGAR, HONG KONG

ARTIST BEGGING, WEST BERLIN

ARTIST BEGGING, PARIS

WOMEN'S ARMY, ISRAEL

WAR MEMORIAL, BULGARIA

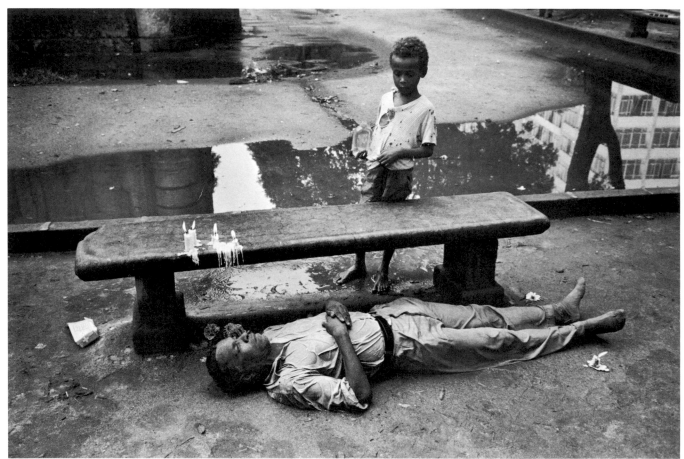

DROWNED MAN, BRAZIL

SLEEPING MEN, NIGERIA

WAITING FOR A MOVIE, CALCUTTA

WAITING FOR TRIAL, U.S.A.

EL SALVADOR

U.S.A.

BRAZIL

LEARNING TO WRITE, HONDURAS

LEARNING TO READ, U.S.A.

NIGERIA

PUERTO RICO

ENGLAND

U.S.A.

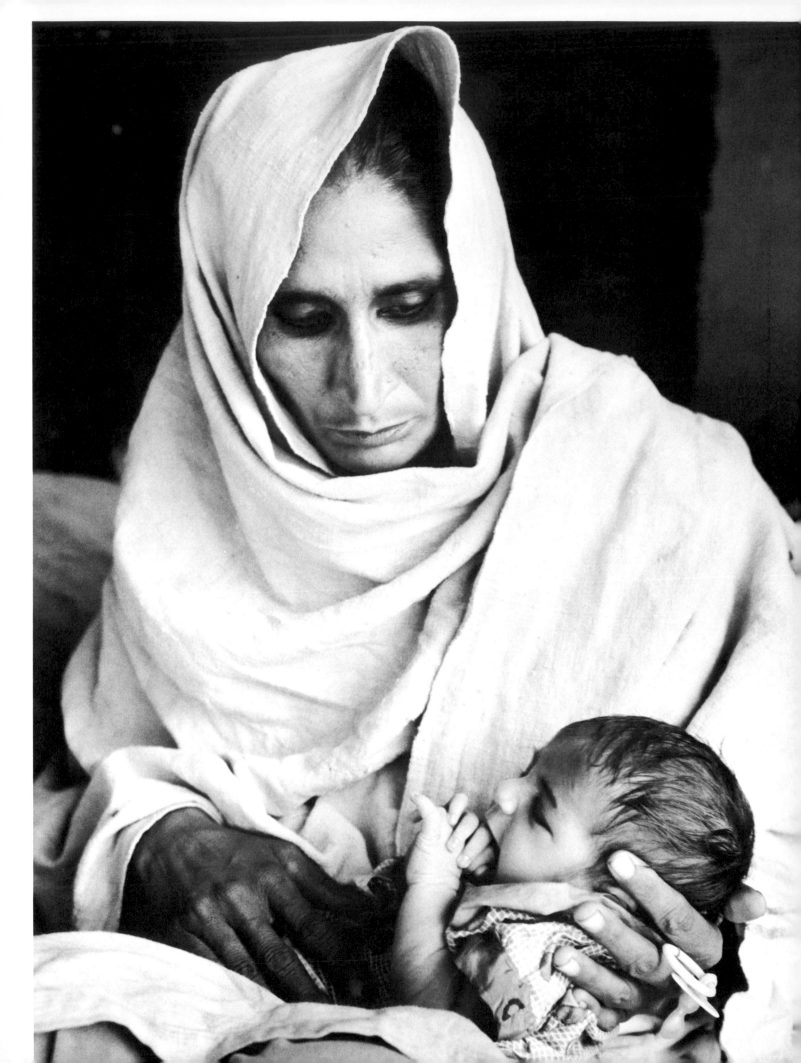

SICKLY MOTHER, PAKISTAN

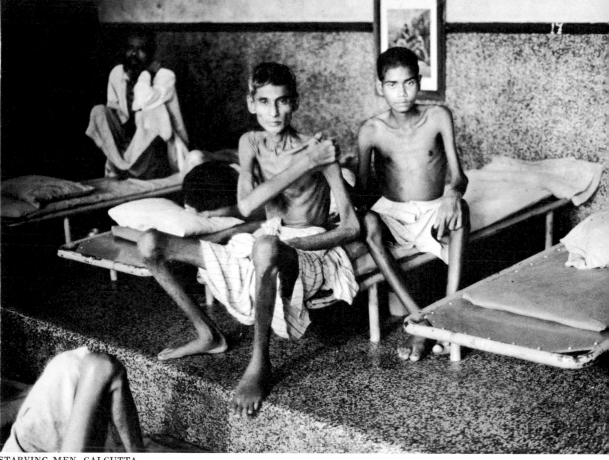

STARVING MEN, CALCUTTA

UNDERNOURISHED WOMAN, HONG KONG

SLEEPING IN A PARK, DELHI

WATCHING SOCCER GAME, CALCUTTA

YOUNG BRIDE, U.S.A.

BARBERSHOP, NIGERIA

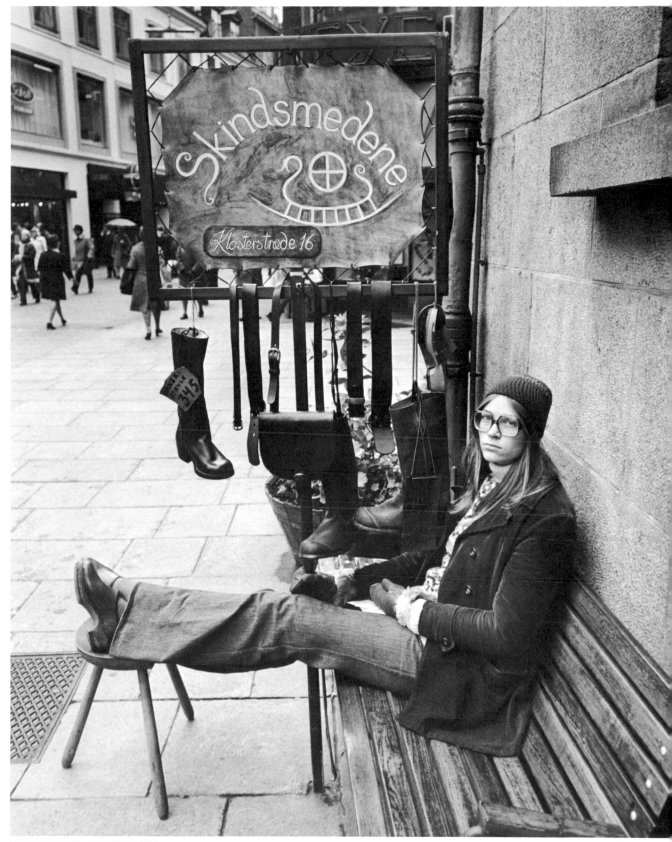

SALESWOMAN, COPENHAGEN

CALCUTTA

The Appearance of the Generation Gap

Doubt mounted slowly through the 1960s as to whether the expansive dreams of a world freed from hunger and want and pain would become a reality — indeed, whether they could become a reality. As United Nations programs designed to alleviate hunger, ill health, and ignorance proliferated, so too did our knowledge of some of the consequences and intractabilities of the attempts to spread modern methods of coping with these age-old problems: schistosomiasis spread by irrigation ditches, a frightening increase in population as public health measures lowered the death rate but did not affect the birthrate, the growth of shantytowns and great proletarized masses in big cities. The enormous increase in suburbs, sterile and far from the heart of the cities, segregated people by age and class and isolated the children who grew up there. The inner cities of the United States crumbled. Paris — so carefully planned as a city of military pride, where triumphant armies could march and rioting mobs could be contained — burst its bounds, expanding into suburban skyscrapers.

And in other areas of life questions were raised. Doubts stirred about sheltering children from the knowledge of death, a practice that had become common in the United States and Britain. There were worries about the way the elderly were being discarded by society, left to molder their lives away in miserable isolation from the rest of life.

But the doubts were still offset by the great sense of achievement: malaria would be wiped out, the great plagues that beset humankind would not come again, whole populations could learn to write almost overnight, countries could be industrialized in less than a generation. Optimism was fed by the beginning of our exploration of space and by the hope that the old quarrels — best represented by that which had led to the trial of Galileo, which made it too dangerous for scientists to announce the burst of a supernova in the heavens — were almost over; thus, in the 1950s, the Pope blessed space research from his papal throne. A big American company made a film that promised that when the resources of the earth were exhausted human beings could simply take off for another planet. The teaching of science in schools in the United States flourished again after decades of neglect, and children's essays reflected this easy optimism: "Without science we would all be living in caves." A decade later, when asked about space, youngsters were still echoing: "I don't see what we are going to the moon for, although I know it will solve the population problem." And after *Apollo IX*, children drew pictures of taxis running back and forth to the moon, obviously a holdover from the anticipations of a decade earlier.

The People's Republic of China published a census in which its government reported "finding" 100,000,000 people who had never been counted before; this was then about half the population of the United States. Miracle corn was developed. Medical students from the developing countries crowded the medical schools, and later, without opportunities at home, filled the more onerous positions in the hospitals of the industrialized world. In the United States, more and more cars poured off the assembly lines, and the material standard of living rose and rose; but, at the same time, workers without education or the hope of high salaries began to refuse to make any effort to work at all. In ancient countries, with centuries-old traditions, the rich got richer and the poor poorer; and in rich countries, which drew on the resources of the whole world, there were terrible pockets of deprivation and misery.

Just as once, at the beginning of the industrial revolution, the craftsman lost his tools and became a dependent proletarian, helpless before the power of organized industry, so the industrialization of agriculture turned the subsistence farmer into a landless worker. Impoverished and starving, the rural poor crowded into the cities. The planners, intent on ideas of organization that neglected the whole person and treated human beings simply as "human components" in an industrial process that might someday need no people at all, neglected the rural regions of the world, from which people were driven relentlessly into cities as refugees from hunger and want. In the cities more skyscrapers were built to contain them; in Hong Kong, the homogeneous Chinese, who had learned to live a crowded life, made this succeed; in oil-prosperous Caracas, where the immigrant millions lived in shanties that they had built for themselves in the shadows of the great skyscrapers, it failed.

And there were struggles for power both within the new nation-states as well as between those supported by Eastern Europe and those supported by Western Europe and the United States. In Japan, on the other hand, industry burgeoned as she entered the race to find outlets for manufactured goods and profits from a favorable balance of trade.

Still, the 1960s began with high hopes. John F. Kennedy, looking far younger than he was, replaced the elderly Eisenhower; the United States challenged the USSR in the race to the moon. In 1963 — right after Dag Hammarskjöld died in the midst of an attempt to settle the disturbances in the Congo — the United Nations held a conference on the Application of Science and Technology for the Benefit of the Less Developed Areas. This was my first experience with a U.N. conference in which the Great Powers debated each other in public and agreed together in private. So the Soviets attacked the Free Enterprise world for advocating population control, declaring, although they themselves rigorously controlled their own population, that the world — we were not yet calling it Planet Earth — could support a much larger population.

The advocates of spreading technology discussed sending obsolete machinery to the developing countries while a few voices in the wilderness suggested designing new, simplified technologies which would be more suitable for the developing peoples; but they were shouted down by both sides, by the leaders of the developing world, who were fascinated with high technology and simply wanted more of it, as well as by the giant contestants for providing it — at the price of political alliance and the provision of raw materials and markets. Doxiadis warned that as many homes would have to be built in the next twenty-five years as had been built in the whole of history. The Great Powers, maintaining an appearance of furious ideological opposition, had agreed that nothing should come out of the conference — and nothing did.

The soaring ambition to make the journey to the moon still held our imagination. But, within groups of those who had organized around the hope of a better world, doubts began to appear in varying forms. Scientists complained that the space effort was producing very narrowly trained young scientists and engineers, and that the rest of science would suffer. An emphasis on aircraft for transport, aircraft for the delivery of troops and war matériel, as well as aircraft to replace railroads or to make travel possible where there were no roads was made at the expense of innovation in the automobile industry. The search for ever more synthetic materials and plastics, and for ever new instant gadgets, preoccupied inventors. The Protestant churches struggled with a belated recognition of class inequality and racial discrimination, as well as with the task of transforming their old missionary zeal into some new form which would recognize the growing independence of those among whom they had previously worked with such an overweening sense of the superiority of everything they had brought with them.

There were occasional bursts of excitement and effort. The Berlin crisis in 1961 occasioned the pathetic and futile command to build air raid shelters and suddenly made people realize that a nuclear attack would differ incomparably from a World War II bombing. The Cuban missile crisis brought home to Americans how very close we lived to reckless confrontation. But spurts of apparent public enlightenment were like fireworks that flare in the sky for only a few seconds. The general feeling of growth, of felicitous change, was too great; there was building, building everywhere. Great skyscrapers went up in London, Paris, Rome, Calcutta, Istanbul. Tokyo was transformed.

Even the persistent and rising voices of the representatives of the new nations did nothing to change the sense of direction that was now shared by the entire world, that the answer to all our ills was quantitative. The need for homes would be answered by more high buildings; the need for food by more tractors, more fertilizer, more pesticides, and more irrigation ditches; the need for education would be met by more schools like those that had been developed in Europe and America a hundred years before; the need for medical care by more hospitals; the need for better maternal care by great lying-in wards which would replace the village midwife and her wisdom, and by forceps used under brilliant lights on women whose position was dictated by the doctor's convenience. A world in which the haves and the have-nots shared the same dream — that the answer could be found in more of everything — was a world in which it was very hard to reverse what seemed like the inexorable spread of a single way of life, a life that would, in the end, benefit everyone.

Then, almost without warning, the recognition of what came to be known as the Generation Gap burst on the world. The first generation of those born and reared since World War II came of age, entered the universities of the world, and began to look around them at a world that had never been seen before, with fresh eyes that had never really known the old world. This freshness of vision evoked in their elders, their teachers, and their parents strange and disturbing echoes. The young seemed to be saying things that had been only half-seen and half-realized by those who had been reared in a quite different world. Perhaps military deterrence was not the answer to the Cold War. Perhaps there were other ways to work out the rivalry between the two Great Powers than generating hostility within the developing nations, or in the Middle East. Perhaps all of our institutions needed overhauling, the Family, the Church, the State, the Economy, political parties, monogamy. Vatican II, initiated by a pope of lowly origin who brought a tremendous and unexpected human empathy to his task, shook the Christian world.

The student rebellion took many forms, many of them trivial, many of them focused on the changing national scene: in the United States, the Vietnam war soon absorbed an interest which had, only a few years earlier, been awakened by the movement to achieve civil rights for Blacks. But even more energy went into the attempt to overhaul archaic university governance. Rules governing the relationships between the sexes were revised and, in the colleges, produced arguments about how small a book could be used to produce the necessary — but minimum — opening of a dormitory bedroom door. In Paris, the great student riots of 1968 included attacks on washing machines as well as the whole of society and — for a brief time — illuminated the responsible planners' hopes. In Pakistan, students rioted against the requirements for degrees. In Japan, students and police faced each other in a confrontation that closed the major universities. We began to realize that we had entered a new era, that young people all around the world shared something with each other, not because they were young but because they had grown up in the same era, exposed to the same world, listening to the same words on their transistor radios, looking at the same images on the television screen, united in their common questioning of the world and the older generations that were responding to these unprecedented events in outworn terms.

The student revolt actually involved relatively few students, sporadically joined by thousands who shared in some temporary episode. Very little damage was done. One computer was destroyed, a few buildings were wrecked; in Paris they tore up paving stones which had been torn up so often before; in California, in Mexico before the Olympics, and in Eastern Europe, they incited ugly, repressive attacks from the police. And, while the questions the young people asked challenged those in power to find answers and cheered aging radicals who had given up hope for a revolution in our time, they also awakened tremendous fears in most of the adult world. Orderly Britain had to deal with thousands of its youngsters pouring across the English Channel, or onto the Isle of Man. Pictures of the great rock-fest at Woodstock were flashed around the world. In the United States, timid old ladies shuddered if they saw a gathering of teen-agers on a street corner.

There was a crescendo of triumphing bigness and venture, confronted by shrill voices raised against it and raised in turn by those who supported it. John F. Kennedy was assassinated, but the Moon Race was won. The American SST, the plane that was to fly three times faster than the speed of sound, the plane whose boom could shatter windows and eardrums alike, was under construction — but never to be finished. The last colonies were being freed and entered the United Nations as small states with few resources, but an equal voice and vote.

And then, just as the student rebellion had burst upon the world, came the unexpected effect of the pictures of Earth sent back from the moon. Those who planned the moon shot had been executing a technological spectacle to dazzle Americans and Soviets alike, and to win applause from an increasingly restless developing world. But the pictures of Earth from the moon produced a different effect. Instead of fueling our determination to master the natural world, it brought home to us, for the first time, how small and vulnerable our planet — the only inhabited planet in this solar system — was. Earth became a little blue ball, a little spinning top, something that human beings could almost hold in their hands, fragile and endangered, something to be cherished instead of exploited.

The environmental movement was born. Like all such sudden awakenings of public consciousness, it emerged out of decades of concern by those who loved the land and worried about soil erosion, those who recognized how vital water was and worried about the sinking water table, those who loved the green earth and worried about the spread of deserts, which had happened so long ago in the Middle East and which was happening again. There were those, like Rachel Carson, who realized that between our monocrops, and our pesticides and artificial fertilizers — which poisoned our lakes and rivers — we were endangering the balance of the natural world and were suffering irreparable losses every year, as species after species disappeared forever, species that had taken millions of years to evolve. But whether the emphasis was on conservation or on preservation, the voices were few, and Rachel Carson was attacked by scientists and industrialists alike. The overall push and direction of human development was toward ever greater mastery of the environment, not respect for or cooperation with it.

The vision of Earth from the moon was just what was needed to trigger concern, the concern that human beings have for that which is vulnerable and small; the concern evoked by the sight of groping day-old kittens still unable to see or by the grasp of an infant's hand around an adult finger. From a certainty that there was a vast reservoir of minerals and fuels, which it would be possible to exploit in ever changing ways, so that a man-made environment could replace the natural world, attention moved to measures to protect what we had received as a heritage from the millions of years in which this planet had been evolving. A spreading fire of enthusiasm was kindled — not yet a very large fire — and one encumbered by romanticisms and preconceptions and ever susceptible to being quenched by attack, by assertions that the young who demonstrated for the environment left terrible litter behind them, that attention was being diverted from the suffering in the crowded ghettos to the well-being of polar bears, that industry would be stifled by new requirements, and that attempts to decrease pollution would harm the developing countries and slow down their progress, leaving them forever trapped in poverty. A fire had been kindled, but it burned flickeringly.

7

A
Corrupted
World?

We are now living in a period of history which to many seems characterized by violence, crime, and a breakdown in morality. Values that people took for granted seem to be in question everywhere; nothing is safe, nothing is sacred. Of course, those with long memories can recall similar periods in their own lifetimes — the tremendous social disorganization that occurred during and after wars and revolutions, the hordes of wild children who roamed Russia, the floods of refugees who crowded the roads of Europe, the massacres that accompanied the separation of India and Pakistan — but all of these were long ago, after World War I or World War II, wars which have receded into history.

And yet these earlier disruptions were quite real at the time. There was the sudden flight from Hungary of the thousands who had waited for years to escape, who filled Vienna for a while with an unmanageable multitude, living in crowded camps where, for example, many saw epileptics in fits for the first time, and were without Dilantin to control them. Or, more recently, there was the terrible photograph of the child in Vietnam, her clothes burned off her back, running and screaming, which has gone to join other pictures that momentarily startled and horrified the world, such as the pictures of the Buddhist monk burning himself to death, the young men in Czechoslovakia and Washington, D.C., who followed him (of whom we have no pictures), the wounded child on the cover of *Life* magazine, and the Parisian weeping as the Nazis entered Paris in World War II. War and violence and suffering in many parts of the world, yes — but vivid as they were, they soon seem to become a part of yesterday.

What remains is a blurred past in which a great many things occurred which appear to have little continuing meaning, unless some relative or friend was involved: British soldiers caught in the internecine fighting in Cyprus, the bombing in New York City in the early 1920s when the freedom of Ireland was at stake, and the bombing in England in the 1970s when the fate of Ireland was again at stake, Jews fleeing from the pogroms in Russia and Poland in the early 1900s, and Jews fleeing from Hitler in the 1930s, Armenians fleeing from the Turks, and Armenians reluctantly returning to what had been Armenia now in the Soviet Union, Ellis Island crowded with immigrants huddled, cold and frightened, Christmas after Christmas, and the Christmas Day when there wasn't a single immigrant there, a television program about the tremendous, growing communities of Greeks in American cities, or Haitians in New York City, or about worshipers of some modern Latin American or Japanese cult in North America.

As with each wave of immigration, among those refugees with historical knowledge or long personal memories, past themes are picked up, so, for me, the assassination of President Kennedy immediately evoked thoughts of the death of Lincoln and the aftermath of trouble that Lincoln might have prevented; but to the young students to whom I talked that week, there was only unhistorical horror that this could happen, now, in America. The deaths of Robert Kennedy and Martin Luther King, Jr., following in quick succession, only heightened the feeling that this time, of all times, had become more evil.

But even if people have some knowledge of past violence — and our history books are filled with the stories of wars and the rumors of wars, with the horrendous atrocities of our enemies and the heroism, in the face of terrible danger, of our own troops — it seems we are much better able to accept the existence of violence due to war than we are to understand the increase of crime. Every time a familiar area of trust breaks down, people in small towns begin to lock the doors, which they firmly believe have never been locked before, and shudder before the thought of a single uncaught villain, kidnapper, rapist, or strangler. To them the world seems more dangerous than it ever was before. And this is especially true when some new form of crime appears, when a group who had been thought to be docile, trustworthy, and sufficiently cowed to be no menace breaks out into rioting or vandalism. Every time a hitherto quiet and protected street is invaded, or whenever country houses of people who have no jewels are robbed, people exclaim: "Not *here!*" "Not *now!*" "Not *my* nephew!" "Not a member of *our* club!" "Not one of *ours!*"

This sense of present shock, so dependent as it is upon our ability to forget all that we have read or heard about other periods — or even just twenty-five years ago — is part of the protective apparatus that makes it possible for us to forget pain. If every pain, every heartbreak, and every fright remained as vivid as when it happened, human beings could hardly survive. Indeed, those for whom past hurt is too vivid become traumatized, sometimes for life; the children who survived the Nazi death camps, living as specially cherished survivors in the safety of postwar England, could be thrown into a panic by the sight of a police dog. But, on the whole, people are able to forget quickly and thus to survive horrors which, when recalled to them much later, may still seem to be quite unbearable.

Groups regather, relatives are found after years of absence and feared death, prisoners of war come home, the survivors of concentration camps resume normal life. Often those who return take up their new lives with seemingly greater strength. This capacity to resume a life that has been rudely interrupted by war or flight is another of our human characteristics which has made survival possible. It involves the ability to accept a changed way of life and, very soon, to take it for granted, indeed, to insist on the correctness of new ways of behavior. Thus, in England in World War II, the blackout became a part of ordinary life, and people quickly learned to measure distances in the dark by walking at an even pace while smoking a cigarette. The reorganization of a disrupted life into a semblance of law and order involves the acceptance of new behavior which would hardly have been thought possible before.

Or consider how the matter of the safety of travel beyond the limits of one's own clan or tribe or village or nation reflects old attitudes and problems as well as new possibilities and problems. Travel away from home has been a perennial problem in human societies, and people have solved it in many different ways — but there have always been times, places, and circumstances where the rules did not apply. In New Guinea, individuals form ties in distant villages which are bequeathed to their sons. When a route is established and becomes firm enough so that one can travel easily along it, then one can sleep in the house of friends with whom one trades and expect to be fed and protected. But sometimes it is not safe to go far from the house of the friend, and sometimes such friendships are betrayed. Among the Eskimo, a stranger might be challenged to a trial of strength, and then, subsequently — if he had arrived alone — be given a wife for the duration of his visit. In many countries, if one has once eaten of a man's salt, one is then entitled to protection; in others, attempting suicide on a host's doorstep establishes a right to his protection. A pilgrim's staff and garb, a beggar's bowl, the practice of an itinerant performance, the jester's cap and bells, the insigne of the king's messenger, all were once the symbols of immunity. At some periods, travel has been possible over wide areas of Europe and Asia, at others, traveling a few miles has been dangerous — highwaymen, thugs, bandits, and despoilers of caravans may beset every mile of what was once a safe, clear road.

In looking back over the history of the safety of travel two things stand out: first, as time has gone on, it has become possible to travel farther and faster, and, second, this extension of travel has paralleled the world's steady evolution toward the interconnected world we have today. In the course of the widening possibilities of travel and trade, as well as of the spread of religious ideas and technological know-how, new hazards in travel have replaced the old. As travel has become faster *and* more extensive, it is hard to judge whether it has become more or less dangerous. Airplanes are safer, mile for mile, than automobiles, but those who travel by air go thousands of miles farther and encounter a new order of risk at their destinations.

The development of insurance for the risks of worldwide travel illustrates another aspect of this question of the safety of travel. In the 1920s, I had great difficulty in getting travel insurance because I was going to what was darkly described, by an insurance employee, as "them interiors." At one point, even in the 1950s, only one insurance company would insure civilians flying in U.S. Air Force planes; but, by the mid-1960s, it was possible to buy insurance for unscheduled flights at nominal cost. For a time, insurance brokers are able to determine the level of risk adequately, and insurance for cargoes that go to the other side of the world gets written. But this may be followed by periods in which no insurance is available for typewriters taken abroad. In some parts of a modern city where theft has been frequent, no insurance is written, just as, at present, insurance companies in the United States cannot cover the possibility of damage from nuclear power plant accidents.

Looked at from the perspective of the distances covered, the numbers of passengers carried, and the costs of travel, as well as of the risks involved, there have been enormous increases in travel in recent decades; but even this growth has been heavily obscured by the speed with which people have accepted the growth in the size and frequency of these as normal. And yet, at the same time, the acceptance of these new conditions as normal is itself obscured by those who persist in believing that every change is of a magnitude never met before and a threat to all that is held sacred. These confused perceptions are further complicated by the export of knowledge and morality from one country to another, of religious and secular missionaries, efficiency experts, technicians, government consultants, military advisers, and agricultural experts.

Each country has its own rules about what is tolerable and what is intolerable. Nothing is as highly variable as local standards of loyalty, of permissible predation, and of corruption. In one country, servants are expected to take part of the market money, in another to get a commission on everything that is bought from certain shops, in another to take home part of the food itself. In other countries, one has to guard against the theft of letters for the purpose of blackmail, as well as the possibility that one's servants are employees or agents of a foreign government, or are using one's cellar as a deposit for drugs, or stolen goods, or merely as scenes for riotous parties over weekends. Each of these will seem terribly wrong to those who follow a different system of social values, of tips, perquisites, gifts, and invitations. And as more and more people have come to live under foreign conditions, more and more often have practices which were tolerated, if not applauded, by those who had grown up with them, been branded as corrupt.

Indeed, one of the most disastrous experiences for a society is the change from viewing certain behavior of its people as tolerable license to viewing the same behavior as genuine corruption and crime through adopting foreign standards. Consider the case of favoritism to relatives. For Americans, civil service examinations and laws against nepotism are supposed to protect society from exploitation, especially the employment of the relatives of individuals with political power. When Americans go to the Philippines, they often express grave moral disapproval of the "corruption" there. If this "corruption" is investigated, however, it is found to rest on several important social conditions: the obligation to assist relatives, in spite of any formal rules to the contrary (which have been imported from the outside), the fact that bribes are necessary in situations where no one would consider them in the United States, and the fact that in the Philippines there is no fixed end to a payoff (in the United States, payoffs are expected to have a fair degree of finality). The codes are different enough so that each appears to the other to reach the heights of "immorality" and to be incomparably worse than that found in the home country.

When the judgmental foreigners also carry with them the idea that theirs is a higher morality, the local people who accept this cultural imperialism and missionary expansion may incorporate into their picture of themselves the accusations of corruption and criminality, and come to think of their own society, not as having the normal, human avenues of evading the law or helping one's friends, but instead as a corrupt society. When the branded practices are not abandoned, when businessmen and politicians continue to favor their relatives, then the society enters a new state of disrespect for itself. Asymmetrical foreign relations may develop between the home and host countries, and the moral infection may go deep, especially affecting the diplomatic relationships between old countries, which are still run on kinship lines, and modern countries, where efficiency is maintained by an insistence on competence. Businessmen and diplomats build up stereotypes of what is required in diplomacy and trade, and each partner in these relationships may adopt the other's bad opinion of his own countrymen, his government's policy of spying, or its tolerance of under-the-counter deals. This in turn may lead to lowered standards of trust in diplomatic negotiation and trade as well as more disorganization and lowered standards of behavior at home.

If we concentrate our attention on one kind of information, then the world will seem to be getting steadily worse; if on a different kind, then it can be seen as developing new ways to handle new problems. Thus, one can drive a car from New York City to Los Angeles with no danger from highwaymen, but one cannot leave a car unattended on the streets of New York City without the risk of the theft of its parts, or of the whole car. The calculation of whether these realities constitute an overall gain or loss in public order is hard to make in any general terms, although this is very easy for New Yorkers — who know it is worse — and the students of highwaymen — who know it is better. That London was an unsafe city in the early part of the nineteenth century and that it is becoming more dangerous today everyone can agree upon, but the perspective of someone who knows what conditions were like before Sir Robert Peel invented his gentle, patient, and unarmed police force of big men is very different from that of someone today who has lived all his life in London presided over by those police, once incorruptible and able to preserve order and enforce the laws against the drug traffic. All through the period since Peel's reforms, there have, of course, been a few criminals in England, well-organized gangs of thieves, parts of London which were dangerous to enter, people on remote coasts who might still rob a wreck, and dock workers one of whose perquisites was theft of cargo. But there was also a hundred years of what seemed to be a very orderly society.

Another way in which people from one society judge another society is by the presence or absence of beggars. Beggars are variously viewed as a sign of a society's poverty or inhumanity or both, a disgrace to a regime, a sign of religious virtue on the part of those who give alms, or a sign of a coming depression. But is a man who draws paintings on the sidewalk and puts his hat beside them a beggar? Is a man who performs tricks for a queue a beggar? Or was the organ-grinder with his monkey — now variously remembered as a symbol of a more innocent age and as a case of cruelty to animals — a beggar? Was he perhaps a now lost felicity of a time when cities had human scale? Or look at our perceptions of poverty in another context: which is more reprehensible, the discomfort of unaccustomed shoes experienced by peasant women in the cities of Portugal, who surreptitiously slip their shoes off when they can, or the failure of welfare agencies in Washington, D.C., to give poor children shoes in time for their entrance into school? In any case, it is certain that the inhabitants of each country would find the practices of the other inhumane.

Even if every social system were, in its own terms, perfect and elicited perfectly moral behavior from all its members — something that no social system has ever been able to do — the contrasts between what one society demands of its members and the behavior demanded by another society can be extreme, and the moral disapproval of each other evoked in two such contrasting societies is often bad for each of them. If either society is covering up, denying, or tolerating behavior which is disapproved of by its own code, the injection of foreign standards into the situation leads to more hypocrisy and self-contempt as a complementary attitude.

Perhaps nothing illustrates these effects of holding mutually incompatible standards more clearly than the question of legitimacy in the countries of the

New World, where Old World insistence on religious marriage and monogamy has been accepted as the norm at the same time that various sorts of unblessed and unlegalized arrangements, as well as the practice of one man being socially responsible for more than one woman, have continued. But neither those who have not married, nor those who have not yet married, nor those who have married can take the same attitude toward the morality of their behavior that their ancestors took before their introduction to Christianity. Both the descendants of African slaves and the descendants of American Indians have, as human groups periodically do, eaten of the fruit of the tree of the knowledge of good and evil, and become conscious of new sins.

So contact between peoples, between different classes, between different ethnic groups, and between nations can produce a sense that the world is becoming steadily less moral, less reliable, less trustworthy and, as a result, give rise to demands for some new form of regulation of behavior. Once pirates and highwaymen became unbearable, legally sanctioned measures were organized against them, and for a time, the high seas and the highways became safe for ship and stagecoach. But today hijackers challenge these norms. When political asylum is given to those who in the earlier period of the nineteenth century would have been treated as criminals in other countries, a new instability sets in, and crime increases. New systems of control have to be devised. Regulatory agencies are set up, police and inspecting forces established, perhaps, in the case of hijacking, to be spearheaded by the associated pilots of the airlines, in an interim period before new international agreements are concluded. And yet as one aspect of air travel is regulated, other practices of evasion may have grown up or the regulatory authorities become corrupt, and when the hijacking is stopped we may find ourselves with a system of organized theft from luggage, such as now happens in many airports.

When a regulatory agency is set up to deal with some new form of lawbreaking that has arisen from some new international instability or circumstance like pioneering some new area, or the establishment of contacts between peoples of different technological levels, the relationship between those who are regulated and those who do the regulating may become the main problem that confronts us. When this becomes the problem, then efforts are made to choose the regulators from the ranks of the regulated, as once the cattle rustler became the sheriff in the American West or the revolutionary augmented the funds of his political party by train robbery before becoming the dictator of a new revolutionary regime. The reformed sinner becomes a saint with greater credibility than if he had never sinned, and the high offices in government agencies are filled with executives from the ranks of big businesses and big unions who once struggled against government regulations.

The questions of morality, of the public observance of declared standards of tolerable and intolerable forms of petty corruption, of the interaction between different systems of morality, loyalty, and rake-off, present an acute problem to our interdependent world in many ways. Social life is based on the trustworthiness of others in the same system. There is no known system of coercion and control that can survive without a large number of shared norms and shared agreements on how people should behave. An erosion of trust produces a condition that is hard to repair; the degree of the erosion defines the limits of possible reform within the system. Once a community has stopped putting money out for the milkman or the breadman or the newsboy because pilfering has started, it is difficult to restore the previous state. When periods of law and order have followed periods of great disorder, it seems to have been because some new set of sanctions, some new apparatus, was devised. Periods of social disorganization invite correction in two directions, one an order imposed by force, which we tend to identify with Mussolini's Fascism, where — as is usually noted — the trains ran on time, the other a radical transformation of society that relies on the creation in individuals of a spirit of deep dedication to new social goals, such as seems finally to have occurred in China after long years of disorganization during which that country was coming to terms with the West.

So the fact — and the sense — of increased crime in the world, of a proportionately larger number of people who are acting for limited, self-serving purposes, often outside the established social system, is serious in that, on the one hand, it contributes to demands for radical change, or, on the other hand, to a sense of hopelessness about the very possibility of change. And when we face, for instance, the possibility of great numbers of nuclear plants, which must be guarded day and night, in all countries, within all systems of morality, within such different limits of tolerable or acceptable deviation, the task of working out real fail-safe devices looms very large. The world has worried along, partly autonomous, partly interdependent, capable of surviving if one part failed, able to survive the perpetual erosion of established systems of morality through the slow development of new systems of morality. But, in the past, the treasures that had to be protected were not uranium, the travelers were not the billions of travelers on Earth, and one failure was by no means fatal, where now it might be.

8

Changing
Expectations

Since human beings emerged as a species, they have been reared in families and have lived their lives within the boundaries of groups to whom they were connected by blood or marriage. In most human societies, most of the time, these are the most abiding and important relationships people ever have. Human beings have also lived in communities, in clusters of families; and within these communities each individual had an identifiable position, kin that each was closest to, neighbors who were more significant than relatives who lived on the other side of the village. There were special relationships that were stressed more than others: brothers who were closer to brothers, sisters to sisters, brothers to sisters, parents to the children of the same or the opposite sex. There might be a recognition in early childhood that one would someday marry the son or daughter of a certain family or lineage. And there might be other socially imposed differentiations, such as people with whom one could not even speak or exchange a glance.

In different places and periods, the way houses were designed and built expressed the organization of family groups in complex ways. There might be separate entrances for men and women, a place where the husband sat and a place where the wife sat, and a special place for a cradle or a cradleboard or a patterned woven mat for the new baby. The wife's tools might be kept in one place, the husband's tools and weapons in another. Sometimes these were thought of as so identified with the users that, when they died, their tools were buried with them. There were places where children's voices were hushed, some part of a house or a surrounding yard might have a particular sacredness; a person might have attained a deep wisdom or a special state, might be a mourner, or a newly initiated youth, or a woman who had just given birth. There were the ways food was stored and prepared, and the regular and expected times and places that people ate together, who ate first, the appropriate blessing or talk or silence that accompanied a meal, the difference between a formal and an informal meal, the greeting and special food offered a stranger.

Sometimes there were no houses; in a dry climate there might be just a hollow in the sand with a light branch bent in an arch serving as the entrance; sometimes hammocks hung from the trees which provided the necessary shelter from the sun, sometimes there was a platform high in a tree. Sometimes houses were built with such care and elaboration that it might take several years to finish one; and fathers planted trees that would someday be the house posts of a son's house. But always there was a pattern, a place for a man and his wife, or, if there was more than one wife or more than one husband, there was always a place where each pair became a couple, eating from the same dish or sleeping by the same fire.

It might take a long time to establish such relationships; children might be betrothed but kept apart, or small girls might be sent to the households of future husbands or allowed to run away for years before the marriage settled down. Often a marriage was made with greater emphasis on the ties between the two families or on the husband finding good hunting partners or lands on which to graze his flocks and herds than upon the personal relationship between the bride and groom.

Marriages might be affectionate or brittle, but they occurred within a community of kin and neighbors. Children's lives contained many other people besides their own parents, or their foster parents, upon whom they could depend. Sometimes preferences within a family group could be expressed, sometimes "a favorite child" could be openly chosen; but more often the patterning within a family fixed definite roles for the eldest, the youngest, or the daughter when there were no sons. And a child might take a long time to settle down after birth; it might be believed that its spirit was still hovering between one world and another, uncertain in which to stay; it might not be given a name for months. And so frequent was the death of children that, whatever the theory of death for adults, the death of children was usually explained differently, as were the deaths of the very old. It was only the death of those who had a firm hold on life which was considered in some way unnatural, or perhaps a supernatural punishment, or the consequence of the malice of kin, or of some neighbor, or of some member of another group who had been angered.

People lived within a known round, winter and summer, spring and autumn, a dry season and a wet season, good years and bad years, drought and flood, years when game and plant life were scarce and others when they were abundant. There were lean years when people went hungry and food had to be conserved carefully, when ceremonies had to be performed to increase the game, or persuade the plants to flourish, or expiate some abhorrent event within the community which was responsible for the evil that had befallen them. The seasons were marked by changes in the positions of the stars — the Pleiades or the morning star — and by the times when honey was plentiful or the nuts of special trees ripened. Survival, health, and contentment depended upon knowing the world that one lived in, knowing how to treat other human beings and what to expect from each, knowing what was owed and what might be asked for, and knowing the idiosyncrasies of each, who was bright and who stupid, who generous and who stingy, whom one could rely upon, or whom to avoid as a working partner or as a companion on a hunt.

Family life depended upon incest rules, rules about when sex relationships were permissible and even enjoined, and when they were absolutely forbidden, as between parents and children and brothers and sisters. The forbiddenness and horror of such relationships was communicated forcefully to small children by tone of voice and gestures of withdrawal. Sometimes these rules of incest were extended to many other people outside the immediate family, to cousins on both sides, to anyone who happened to live in the same household, to everyone who lived in the same village, to members of large clans living apart, to larger groups who kept the same taboos, or acknowledged a totem relationship to the same animal or plant. Sometimes sex relationships were limited to one person at a time, to a spouse of the opposite sex; sometimes they included a wife's sisters or a husband's brothers, or a wider group during holiday seasons, or special people who became initiators. But there were always rules, and always enough people who did not, or could not, learn or keep the rules, so that the knowledge of the rules was kept conscious, although often just on the fringe of awareness, only to come to full consciousness when they were breached, as nonliterate people who have never heard of grammar still rebuke a child who makes an error in speaking.

As the central core of the household, parents and children were protected in such a way that children could grow up to express physical closeness without a fear of premature sexual exploitation from those who were stronger or older. So also the boundaries of the larger group, among whom sex partners could be found, were fixed; there were people whom one did not marry, people with whom close contact was regarded with a fear often tinged with the same horror that was associated with incest, people who were seen as not quite human beings, as like animals with monstrous and frightening habits. And there were taboos against sexual contact with animals, with people in special states, like menstruating women, hunters before they left for a hunt and warriors before they left for a raid, people who had touched a corpse, the corpse itself, and with those who had shed blood. The permission for sex and the prohibition of sex and other contacts that might lead to sex — looking at or talking with someone, touching, calling by name — all were patterned.

Just as the group of people with whom sex is totally forbidden may be limited to parents and children or widened to include several hundred more distant kin and fictional relatives, like godparents or trade friends, so also the place where the outside circle is drawn, beyond which one cannot go, widens and shrinks among different peoples and under different circumstances. The line may be drawn in terms of language, skin color, occupation, or religious belief. People who have come from an open society in which, theoretically, at least, there are thousands of suitable mates, may join a religious cult or a small political group of a few hundred people, and the number of possible mates will shrink to very few. But whether the core incest group and the group within which one may marry are wide or narrow, the boundaries share some of the same feelings of horror. If the horror of marrying the stranger, the nonbeliever, the uncommitted, the pariah, or the commoner becomes too strong, then there is likely to be a relaxation of restrictions in the inner core. Thus we find incest is frequent among isolated groups who are hostile to all outsiders or among royalty, where there are no others considered "fit" to marry.

Tightening the outside boundaries creates groups that have great internal coherence, but this often arouses antagonism among others who do not have equally strong taboos against outsiders. As long as Christians, Moslems, and Jews equally opposed intermarriage, as long as the different castes in India agreed upon the necessity of maintaining caste rules against intermarriage, large societies could exist in relative peace, just as small, primitive communities have marriage taboos within clans or between groups that are too closely related. A few individuals, more passionate or individualistic or weaker than the others, will break the rules, and the punishments meted out to them, death, exile, or excommunication, either strengthen those who do not break them or pave the way for changes. Societies can exist for relatively long periods of time, even with many people flouting the rules which govern relationships between the sexes, but we have as yet no reason to believe that they can do without them.

If a people have discovered substances that intoxicate or soothe, we find that their use is also patterned, that there are times when wine (or beer, or distilled liquor, or stimulating, intoxicating, or soporific drugs) should or should not be used, that tobacco was to be smoked ceremonially in peace pipes or at mourning feasts, that wine was drunk to celebrate special occasions. The interplay of the sacred and significant with the secular and profane gave a rhythm to social life, just as the processes of birth and death, and of mating and bereavement, provided for alternations of hope and despair, joy and sorrow, contentment and bitter disappointment.

Peoples vary as to where they locate the period when things were better than they are or worse than they are. Some peoples look backward to a lost paradise where food could be obtained without labor and everyone lived in a world of peace and plenty. Others see the present as a recently attained improvement on the hazards or hardships of the past. Still others are sustained through this vale of tears by the hope of a better life in the future or in the hereafter. These stylizations of expectations may take the form of a mythical golden age, or of a recent past, "When we lived in the old place," or of a glorification of the joys of childhood or early youth. They may become the celebration of a newly attained prosperity for immigrants flourishing in a land of plenty, where, even though the streets are not paved with gold, there is more to eat and greater security than they had known before. They may express the pleasures of the married state for young people who have finished with the rigors of the parental home or the demands of initiation or schooling. The sustaining hope of a better future may be expressed in the hope of a serene old age, or of a better life for one's children, or of a better world, while the hope of an afterlife is expressed in images of a release from care and work and the reuniting of the separated.

Each of these ways of viewing what is best in the human condition is open to everyone no matter how sedulously a particular culture or a particular religion may emphasize one at the expense of the others. The exigencies of individual life histories, of differences in temperament, and of the availability to everyone of the experiences of maturing and aging make all of them continuing alternative ways of viewing the world. Among people who have only an oral tradition, it is very easy to shift such emphases; the editing of tales told many times goes on continuously, and the past can be revised without anyone being conscious of it happening. An unexpected view or a different way of doing things may devalue the past as well as the present almost overnight. Indeed, old ways may not only be devalued but completely forgotten or assimilated to the present, and the present may come to be seen as the only imaginable future.

In the literate, rapidly changing, interdependent world of today, whole populations may find their moods oscillating between hope and despair or temporarily absorbed in some overwhelming obsession or fad. But it is the ever availability of biological rhythms that underlies the human responses to cosmic conditions and events. The models of birth and growth, maturation and mating, aging and death, male and female, parents and children, kin and neighbors and friends, of the small community which is too close and the larger community where relationships become too fragmented, formalized, and distant, all are always available, ready to provide the framework for a different phrasing of experience.

The extent to which humankind must rely on the recurrent and given conditions of human life and the extent to which progress and hope depend upon transcending them is one of the issues that will have to be argued out not only in the next quarter-century but presumably as long as human beings exist, whether human life is continued on this planet, on artificial worlds, or on some distant, not-yet-explored but habitable star. One of the problems that confronts us is the determination of how many of the changes which have occurred in our long history are irreversible, how many still contain alternatives recurrently open to us, and whether we will choose to dwell in some real or imagined golden age of the past — when the snowdrifts were higher and the flowers smelled sweeter in the springtime — in the actuality of the present, or in a utopian or realistically obtainable future. An equally intractable characteristic of the universally given conditions of human existence is the fact that the past and the future are much easier to transform, by idealizing or diabolizing, than is the present. A whole nation may live upon a myth of former days, when its armies were always victorious, its princesses beautiful, and its kings benevolent, although none of these mirror any past reality at all. Such dreams are, of course, vulnerable to attacks from others whose own myths of past glory contradict such tales; hence, nations with unreal myths of the past may seek to restrict their contacts with possible critics. Similarly, if the hopes which must attend any revolution are to be maintained, the further their realization is put in the future and the firmer the insistence is that it is *future, true* egalitarianism, or harmony, which is being sought, the easier it will be to keep people working for the new shared goals.

It is the present that is intractable — the bitter actuality of no food, of no water, or of no roof over the heads of one's children — both for those who once had food, who now see their own children starving in the midst of plenty, and for those who are expected to accept the meager present in order to attain a better future. But it is the present that is the hardest to manage, when the descendants of kings who drank elixir from goblets must drink contaminated water from a chipped cup. Of course, the chosen of the Lord for whom seats are reserved in Heaven can put up with almost any hardship here on Earth, if they see themselves as mere sojourners on the way to a celestial reward. But for those who have only their earthly years to enjoy living, a life which is sustained by no vision of either the past or the future, the harsh realities of the contrast have far more impact. If they are still living in dependence on the natural world, on the spawning of fish, and on the rains that come or do not come, the present varies independently of their wishes. There is always the possibility that the next year will be better; but however complicated the ceremonies of magic or prayer with which they propitiate or attempt to modify the world around them, the actuality is always there. Sometime it will rain, and, someday the rain will stop. The specter of a time when the rain will not stop lies upon the imagination of the people of the world in the legend of the Flood; but for the life of everyday, when that life is one which is dependent upon nature rather than dependent upon the ways in which modern man has mastered nature, the present lacks uniformity whether of promise or of disappointment.

This is still the way life looks to most of the peoples of the world when they think of the present. Whether it is good or bad, it is out of their control — but it will change, for nothing lasts forever, and

> *We thank with brief thanksgiving*
> *Whatever gods may be*
> *That no man lives forever;*
> *That dead men rise up never;*
> *That even the weariest river*
> *Winds somewhere safe to sea.*

Or as Ruth Benedict used to say to those who expressed impatience with some situation: "In the end, people always die or move away."

More serious problems arise when people enter a money economy and adopt the belief that if one has money one can purchase everything that is needed or desired: food, water, shelter, wine, travel, pleasure, and relief from pain. Thus water was twenty shillings for a quart on the goldfields of Australia, but those who struck it rich could afford it. Even in the conditions of greatest famine and plague, flood and earthquake, and war and revolution, it is those with money who can escape and those without it who are left behind. After the last earthquake in Honduras, the business offices and the homes of the wealthy were not restored at the dangerous, thrice-destroyed site, but the homes of the poor will be rebuilt there. The wealthy whose children are born with some rare disease can travel to a specialist; the wealthy can pay for other women to suckle their children; the wealthy can pay for rare imported foods. When people move from a world in which the few rich and the many poor were alike subject to the vagaries of the natural world, in which only a small surplus could be accumulated by the wealthy and lucky and provident, to an urban life in which money is the universal medium of exchange, everything changes.

There are still many poor urban people who deal with the problem of money as if it were completely beyond their control, a matter of good or bad luck, of the whimsies of fate. They buy expensive food when they have money, and live on beans when they have none; their supplies are augmented when some member of the family or the small group happens to get a job, or happens to win on the horses. A run of luck will certainly change and change no matter what a person does; so the presence or absence of money is treated like the occurrence of good or bad seasons: there is no garden to be cultivated diligently, only trees from which the fruit can be gathered whenever the fruit happens to ripen. The urban poor, especially those recently come from the country, have less control over their lives than they had as an agricultural or herding people but they can move, go back to the country, or go to a larger city; they can be alert for opportunities, they can join a thousand unemployed men to shovel snow or pile sandbags along the bank of a flooding river; or they can ransack a wreck on the beach or an overturned cart of vegetables. They are there if a riot starts and stores are looted, but they are not, as a rule, revolutionaries, for they, like their fellow countrymen and -women whom they have left behind, have no sense that they live in a world in which their own efforts can change the world. They may join small cults or churches with a pentecostal message, rely on amulets and charms, gamble whenever they have money, all very much in the spirit of the world from which they have so recently come, where all had to adjust to the weather, bear the bad times, and seize the good.

Today, the sense that everything can be bought for money — a pervasive characteristic of a money economy — has been combined with a spreading egalitarian ethic that everyone deserves the best that is available anywhere. In the past, those who were spiritually affronted by the terrible contrasts between rich and poor and who took seriously the admonitions, "It is easier for a camel to go through the eye of a needle, than for a rich man to enter the kingdom of God" and ". . . go and sell that thou hast, and give to the poor . . . ," gave to the poor, or joined the poor, giving up wealth and riches, stripping themselves of worldly goods, living a life of voluntary poverty and asceticism. Today, although there are traces of such voluntary simplicity among some of the young of the affluent societies, the overwhelming emphasis has been not in this direction but in the development of the view that those poor by birth, position, lack of education, or residence in an overcrowded, impoverished part of the world are now entitled, as human beings, to whatever affluence they see on television or in picture magazines. In the past, the servants of the great houses, who polished the silver and fluted the curtains and washed the crystals on the chandeliers, returned to their own dwellings, unadorned hovels whose misery was relieved only by the scraps that fell from their mistresses' tables. Today, the poor see in magazines, in films, and on television people who appear to be little different from themselves, living in splendor while they live in squalor. The belief that people were born into a given station in life and that certain amenities and luxuries were the proper prerogatives of the rich, the upper caste, the uniquely successful, while others should serve the Lord in whichever station in life they had been called to, is disappearing. It would have been inconceivable forty years ago that a poor boy who had assaulted and robbed a stranger on the street should have said on television: "But I had to have money. I couldn't wear the same clothes every day."

It is these changes in expectations that require changes in the life-styles of the industrialized world, where many are now living at a level which cannot be provided for the poor even in the affluent societies, still less the poor in the developing countries, without irreversibly depleting the resources of the planet.

IV

BALI

BEGINNING
AGAIN

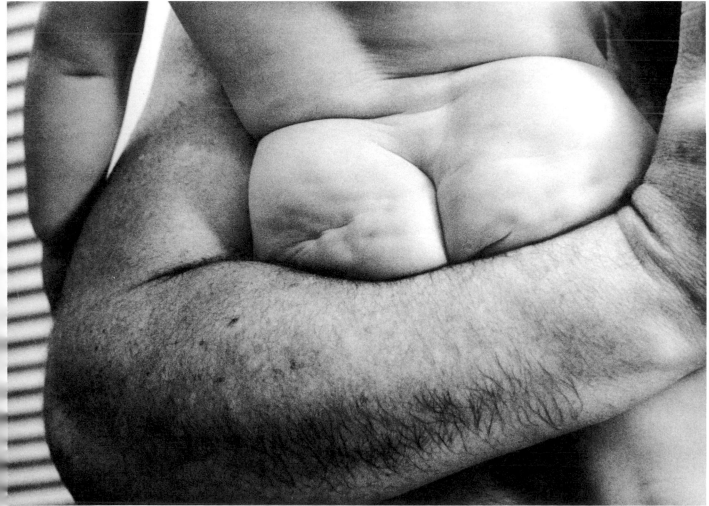

BOY AND HYDRANT, U.S.A.   FATHER AND CHILD, U.S.A.

DIVING, EL SALVADOR

BATHING, BALI

U.S.A.

BALI

SICILY

EMBRACING, SPAIN

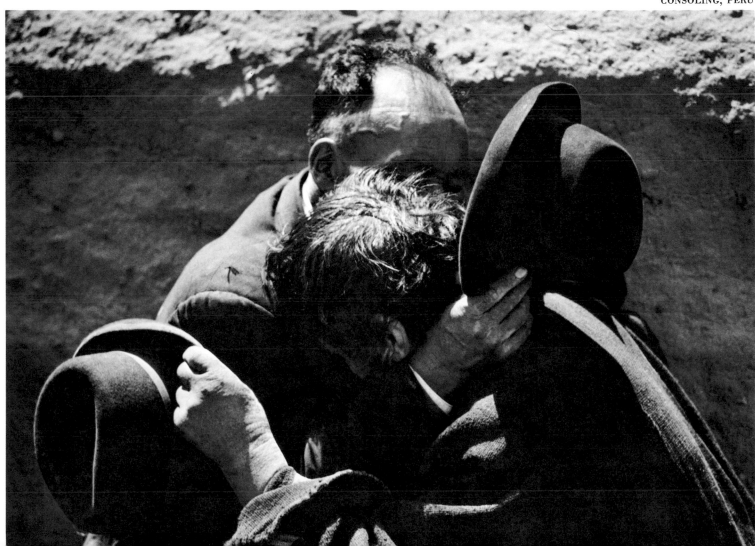

ENCOUNTER GROUP, U.S.A.

WEDDING PARTY, MOSCOW

MOTHER AND CHILD, ITALY

BEER HALL, WEST GERMANY

TUG-OF-WAR, U.S.A.

WRESTLING, U.S.A.

NIGERIA

PULLING NETS, PANAMA

CHOPPING WOOD, EL SALVADOR

SPINNING WOOL, PERU

PLOWING WITH OXEN, BALI   SIMULATED MOONSCAPE, U.S.A.

THRESHING WHEAT, GUATEMALA

STREET SCENE, U.S.A.

PEDESTRIAN MALL, SWEDEN

SPACE CENTER, U.S.A.

SCULPTOR, FINLAND

COLLEGE DORM, U.S.A.

INSIDE A CAMPER, U.S.A.

ORTHODOX SCHOOL, ISRAEL
SUNDAY AFTERNOON, NEW YORK CITY

TEMPLE, TOKYO

RICE PADDIES, BALI

BEACH FRONT, PORTUGAL

ACTRESS, TOKYO

CITY PLANNER, SYDNEY

PHOTOGRAPHER, INDIA

STUDENT, NEW YORK CITY

SANDWICH VENDOR, FRANCE

LOW TIDE, BALI

9

A Fuller
Consciousness

One way we can understand the human past is to view it as the history of our increasing consciousness both of ourselves and of the natural world of which we are a part, as the history of our increasing awareness of the self and our increasing understanding of the universe within which each self exists and has significance. We can think of the evolution of life on this planet as a process within which life itself, as evolved and manifested in human beings, has taken over the direction of its own course. In such terms, we can think of ourselves as on something like a man-made spaceship, or we can recognize that we live within a biological system and that, however much we may probe its nature and systematize our understanding of it, however much we imitate it and improve upon it, we cannot claim credit for bringing it into existence. We can treat this Planet Earth like an expendable machine, or we can recognize that our dignity is dependent upon the respect we pay to the origins of our earthly life and upon the responsibility we take for its preservation.

Each evolutionary step forward has carried new possibilities and hazards: with the development of language it became possible to think about people at a distance, to fear and hate them or to identify with them, and to devise large systems of government within which all might cooperate with one another. Script and science made it possible to magnify the scale of our activities and to devise the institution of warfare, until a stroke of the pen could turn a hundred million people from fellow human beings, whom one must not kill, into enemies of all that we hold dear, whom it was our duty to kill. As we developed weaponry from the crossbow that could slay one man at a distance, to the battering ram that could break down a fortress gate, to a bomb that could disintegrate a city, to plans for a personalized laser weapon that could melt a human being to nothingness, we also developed ships and planes that could rush food and medicine to scenes of a disaster. The automobile traffic within which children are run over and the ambulance that rushes the hurt child to the hospital are part of the same system, both expressing what human beings have been able to accomplish for themselves in a relatively short period of time, after millions of years of slow development.

There are those who believe that it has been the pursuit of limited purposes — building a city, defeating an enemy, designing an airplane — without paying attention to the whole of life that has led us to the current point of extreme danger, and that our only recourse is to return to a way of life in which people, living simply in small groups, listen to the rhythms of their own hearts as unreflective parts of the natural world. A hundred years ago, this was a counsel which might have been accepted by the most sentient creatures in the solar system, who might then have gone back into caves and hermitages, abandoned the cities to predators, and sought for fulfillment in meditation and prayer. Then, the population of the planet was only one billion souls, there were still vast areas of unused land which would not be destroyed by the scattered users, and there was wood to burn and coal to mine. During World War II, there were still those who predicted another Dark Ages, when the records of human achievement would have to be hidden in the ground to be disinterred by future generations more able to use them than we. There are still many who take this point of view and suggest that we abandon our attempts to use technology for the improvement of human well-being because all such attempts of limited scope and purpose are bound to fail. As we drive, with however great strength and determination, toward any human goal, we inevitably involve, they say, not only ourselves but the earth on which we live — of which we are only one species — in imminent disaster.

Four hundred years ago, the position that Christians should save their own souls while society and its institutions collapsed, and the rest of mankind was hurled to destruction, presented, at least as far as those who believed it knew, some possibility of fulfillment. Great civilizations had disintegrated before (only the lion and the lizard lived on in the lands where kings had reigned); they had done so before, they could do so again. Since the invention of the bomb, however, we have begun to realize that there is no road back to earlier simplicities, away from the complexities which our pursuit of limited goals has generated for us. I remember how struck I was the first time that I heard the statement that, if human civilization were to begin again now, it could not take the same course — too many of the resources on which civilization has been built have been destroyed. And, indeed, we clearly cannot go back to an earlier stage and begin over again, weave our own cloth and grow everything we eat, build our own houses and each instruct our children in the few necessary things that they should know.

For Americans, retiring into isolation, using technology to surround us with missiles, is just a way of attempting to go back to a previous state of immunity from attack. But to go into an isolation that once was protected by wide oceans is no longer possible in a world where spy satellites, and weather and resource satellites, pass back and forth over our heads. Triage, which assumes that we are a handful of desperate people in a lifeboat rather than members of a world community, is not a morally acceptable solution. And communications are too fast, the weapons delivery systems too effective, and the methods of biological and chemical warfare too certain to kill and too difficult to control once let loose.

At this moment in history, if we ever hope to have a choice again, we have to recognize that we do *not* have a choice between going back to a preindustrial state or going forward. But we *can* go back twenty-five years to the point where we took a wrong turning and start over again, still on the same main path, but this time seeing the world as a whole, in all of its diversity, part of nature, not set against nature. This would be going forward, but facing all that we have done and all that we have left undone that we should have done, forward to a conscious purpose that is wide enough to embrace the safety of the life of a solar system within which we are the only self-conscious beings.

It is open to us to correct the damage that has been done when we gutted the resources of the earth, polluted its skies, distorted the lives of millions of its inhabitants. It is open to us to correct for the consequences of our limited purposes and partial understandings by widening our purposes until they coincide with the solar system within which we live. It is open to us to focus our purposes until they include each living being, those now alive and those who will be born in generations to come.

We have the means to do this, the kinds of knowledge and the instruments and things to do it, computers that can calculate fast enough so that we do not have to wait years to understand what has happened or what the consequences of actions taken now will be. As we became human when we could think of, and talk of, and take into account those whom we had never seen, we can become more human because we can now think of the fates of all the peoples on this earth, at once, as part of one system. We are not limited to thinking of them as a thousand fragmented bits, fragmented not only in time and space, but fragmented by our previously deficient methods of thinking about them as primarily physical or biochemical creatures. As schoolchildren we read that Caesar had to do everything at once, and Caesar was the conqueror of only a small part of the world. If we can invent a "macroscope" for the world, see the world as a whole, and act in the knowledge that it is one, we can keep it whole.

10

Complementary
Models

People all over the world have two models of what life may be. One of them is the model of a world filled with living things that grow without interference as they have grown for a million years: trees in the forest, grass on the hillsides, wild plants and wild creatures, living in a relationship of precarious and changing balance with one another. The bodies of women have provided for human beings an epitome of such natural rhythms; thus, the infant grows, unseen and undisturbed, cradled safely in a protective fluid inside a veil within its mother's body. This persistent task of women has been essentially unmodified throughout human history, until this century, when the busy, inventive minds of men, ever concerned with wresting its secrets from nature, began to change our conception of pregnancy and birth from something that belonged primarily to women, and to women attended by women, to something that would fit with the other model of life, a man-made and man-controlled world.

This building of a man-made world is a process that has been going on for a long time, since the invention of horticulture and herding, when human beings learned how to domesticate and breed animals, and how to save seeds and plant them in the ground. Later men began engineering and architecture, and navigation and astronomy were developed. As the mastery of men over nature developed, women lived in the tents and mud houses, the teepees and terraced pueblos, and turned them into safe places where children could be fed and men welcomed from the chase, or the war party, or a long journey of exploration. When women belonged to a herding people that traveled swiftly, they prayed that their babies would be born at night when there would be time to rest. But whether the towns and camps designed by men were safe or dangerous, whether the society was simple or highly stratified, whether pantheons contained goddesses or whether there were occasional queens whose royal blood was more important than their sex, the place of women remained substantially the same, submitting to the natural rhythms of their child-producing bodies, while men, taking their cues from the women who reared them, believed that it was their task to go out and achieve in the wider world. In such a context, childlessness, for which there was no artificial remedy, was branded as unnatural, and even the happiness of kings and queens was marred if they failed to have children.

Both men and women were presented, generation after generation, with a world in which life followed its own laws and a world that was made by men who studied those laws. There were birds that flew and fish that swam, and, in time, airplanes and ships that imitated their capacities. There were plants that thrived on sunlight, and there were fields turned into factories where grain was grown according to plan. The arts and religions could use either model; a poem could be seen as something that grew as naturally and effortlessly as a child or as something that was made, according to rules, as precisely as a man-made tool. A pot could be produced by a potter at his whirling wheel or by a factory worker running complex modern machinery. The stars and comets in the sky could be watched with wonder, but their paths could be predicted. Fireworks and skywriting planes and the lights of *Echo* circling the earth went to join the stars in the near sky. Nature as a model, as a surround to be respected and cherished, or nature as something to be studied only to be improved upon in a man-built world, these were the alternatives, sometimes seen as complementary, sometimes in opposition, but always as possibilities.

Today, for the first time, the contrast between the natural rhythms of motherhood and the constructed roles of fatherhood is being challenged. Artificial insemination can substitute now for intercourse; intercourse and the conception of a child can be separated, not by voluntary celibacy or clumsy mechanical barriers, but by direct interference with the body chemistry of women; menstruation can be stopped completely or prolonged; and infants born ever more prematurely can be kept alive outside the womb. Paralleling this greater control over the reproductive processes of women's bodies has come the control of disease and the possibility of bringing plenty to whole populations. With this, the need to bear many children so that a few will survive is being replaced by the admonition to bear fewer children because those who are born will live.

We can see all these changes as freeing both men and women from the burdens of parenthood, as societies in the past have been willing to do when they perceived that their population was growing too fast. Where once we had the voluntary celibacy of the monk and nun, the mendicant, the armies who wandered from land to land, the men who left home when their first child was born, we now have artificial methods which separate sexual intercourse from procreation. Today, the bodies of women are changing from something natural, to which men and women had to submit, into something like a machine whose timing and activities can be deliberately controlled.

Through the ages, while the relationships between men and women were developed, the roles assigned to them by society and the strains and discrepancies within those roles have come about because no ascription of one kind of behavior to one sex completely excluded the possibility of that behavior occurring in the other sex — indeed, there has often been tremendous overlapping. Among very simple peoples, differences in personality loom very large, and peoples resign themselves to the existence of manly women and womanly men no matter how rigid their division of labor, or contrasts in clothing, or the importance assigned to one sex or the other. But in larger societies, the assignment of sex roles may become stifling if men and women have narrow, socially enjoined roles. For males, parenthood is often seen as a narrowing of their role in life; but male children can be offered a great variety of life choices which are not primarily ways of supporting women and children. For females, motherhood, which is usually an existence that is less specialized, more confining, but also more preservative of the whole mystery of life than the male role, need not become the complete preoccupation of a whole lifetime in the creation and care of individual human beings.

Before the invention of the X ray and other techniques of research and diagnosis, the process of pregnancy and childbirth remained shrouded in mystery, and neither men nor women knew much about the procreative functioning of their bodies — although this knowledge diminished and increased over time: hunters know more than agriculturalists, herders more than farmers, cannibals and professors of anatomy who teach medicine through autopsies on dead bodies more than those who treat the bodies of the dead as sacred and inviolable. But the knowledge that we have today — dramatized as it is by the legal controversy as to whether or not women's examination of their own bodies is "practicing medicine without a license" — is of a different sort than the knowledge that had been gained before. There are changing demands at many levels among those who wish to live in a world where nature is respected, where human beings can learn to listen to their own hearts, to follow the tides and the seasons, and to attend to the miracles of birth and growth. People want to see the creation of any material thing as part of the naturalness of life, a tool as an extended hand, flight as the realization of a human dream:

> *Desire it is that flies; then wings are freight*
> *That only bear the feathered heart no weight.*

But, there are also those who are transported by the possibilities of miniaturizing electronic circuits and machinery as is done for space flight, of going beyond fission to fusion, of producing human beings without the act of sex, of producing an individual from a small tissue sample of another, and of altering human genes. Forty years ago, Lawrence Frank said that the greatest ethical issue of our time would be birth control, anticipating the myriad ways human intervention in life processes would present us with an increasingly serious moral dilemma. Man-made interventions in agriculture and herding, as well as machine manufacturing, increased speed of transportation, pesticides, fertilizers, hybridization, automation, and artificial insemination, have each in turn transformed the world we live in to the extent that there is now a sense of mounting catastrophe. There is the sense that the intervention has been too great, too rapid, and too unthinking. Thus, an animal, an insect, or a plant is transported from one part of the world to another, and turns into a pest or a weed. We note, without yet sufficient willingness to take enough possibilities into account, the multiple and interrelated ramifications of introducing some new artificial process, and try to grasp the lessons.

Or consider a new invention such as bottle-feeding with animal milk. This saves the lives of many babies who would otherwise die, but animal milk is not a perfect substitute for mother's milk and some babies still die. When there is a move to reintroduce breast-feeding along with bottle-feeding, still fewer babies die. Formulas are developed and improved as our knowledge of nutrients increases; the dangers of using animal milk are overcome by pasteurization and refrigeration. Formulas and breast-feeding can be prescribed for the needs of particular babies when the physician cannot control the moods of the mother as easily as he can the proportions of the formula. But milk production in sufficient quantities is impractical in many parts of the world, so powdered milk is manufactured in industrialized countries at factorylike dairies where the employees milk cows they do not know as individuals to be treated with care but as anonymous creatures standing in serried ranks. The employees become manipulators of gauges, hands no longer skilled in stripping the udders of animals they have known since they were bumbling little calves. The powdered milk is then sent far away to be fed to the infants who are born in such large numbers in the shantytowns of the new cities. But again there is a problem of cleanliness and many babies die. Sometimes the supply of milk fails altogether when war or some natural disaster separates mothers and their babies from the source, which may be on the other side of the globe.

In the world today, more and more infants are severely undernourished, and yet live, frequently impaired for life. And then, wise men and women, watching the infants who die and the infants who live but as impaired human beings, say, "Let's go back to breast-feeding, back to the natural process by which human infants have always been fed, where there are no interruptions in transportation and no contamination of the milk." But when we go back to breast-feeding, often in times of scarcity and danger, new problems arise. The longer that a population has depended upon the artificial feeding of infants, the more likely we are to find anomalies in mothers and children, to have mothers who cannot produce enough milk or who cannot produce milk of the right quality, or infants who have trouble sucking.

Here again there is the need for intervention; sensitive therapists make films — with all the apparatus, the lights and video cameras, available in the industrialized world — of new mothers and their babies to study how a mother and a baby who don't "fit" can learn to adjust to one another. Where in the past, and in many parts of the world today, an infant would have died, now, we can help it achieve a better adjustment to its environment and survive. At each stage in the process in which human ingenuity has intervened in the relationship of human beings to their own bodies, new conditions and problems have been created which do not have the kinds of unplanned solutions upon which human survival has depended in the past. Then, the most fit lived and the less fit died, or there was the kind of intervention which, interestingly enough, we now brand as "unnatural": infanticide, or feeding the stronger child so that it may live instead of the weaker younger child, or the gradual withdrawal of the breast-feeding mother from a child who is losing weight in her arms.

All over the world, a battle is going on between older forms of behavior, once adequate to keep the population of small groups in balance, and the introduction of practices which interfere with such adjustments. Frequently these new practices stem from decisions made on the other side of the world, often with the result that nations are modernized in ways that make their cities completely unmanageable. There are battles over the acceptance of contraceptives, the introduction of new foods, the legalization and spread of abortion, which has always been the last resort, short of suicide, of women desperate from an unwanted pregnancy. Thus, many women see the possibility of abortion or contraception as providing a better chance for the children they already have, and some women see these methods as ways of escaping altogether from the burdens of maternity. But there are men who oppose abortion as a signal of their respect for life, which, however, they may denigrate every day in their advocacy of warfare, in their destruction of the environment, and in the hazards of all sorts they impose on their fellowmen. And there are men who wish to make their own, to subject to their will, and to manage and construct the whole process of childbearing — artificial insemination, artificial implantation, gestation outside the body — so that, apparently, men and women will at last have an equal — and an equally unnatural — share in the production of the next generation.

Periodically, the unending controversy of how much human beings dare to intervene in the whole balance of life, how much they can alter the interdependence of living things, without destroying themselves and the rest of the living world at the same time, becomes a crisis. We become fascinated with heart transplants — and, incidentally, have to deal with the willingness of a white man (who has lived in a society built on the subjugation of black-skinned people) to accept a black man's heart, just as in World War II, the lives of soldiers were sometimes saved by the blood of those whom they had set out to exterminate. Each extraordinary act of intervention, possible only with the highest medical technology and the highest development of surgical skill, carries with it a restatement of this human dilemma.

The controversy, which assumes very sophisticated forms in the theological disquisitions of the priesthoods of the world, and which confronts every villager to whom the "benefits" of modernization have been brought, is bound to continue as long as we remain human beings and do not abdicate entirely after constructing machines that can take our place. For most of the peoples of the world, there is still a sense of closeness to their own bodies and to the natural world, and although it is permeated by cultural conventions, it, nevertheless, is real, immediate, and biological. The new baby may be powdered or greased with products of various degrees of naturalness — bear fat, butter, fuller's earth, medicated talcum powder imported in plastic containers from ten thousand miles away — but the women still breast-feed their babies. Death may come from the malfunctioning of new machinery, or from eating seed grain that has been coated with a poisonous substance to prevent fungus, or from eating fish poisoned by some nuclear accident at sea a thousand miles away, but those with whom they have lived in the closeness of kin and neighborhood still close the eyes of the dead and follow their bodies to the grave, lament the babies who die, and hope for more babies who will live.

As people struggle to adapt to the differences between the traditional meaning of life and the mechanized life-style that is overtaking them, they increasingly demand a share in deciding what form the new is to take. The people of each country newly urbanized, each ancient civilization where millions have lived with unpaved streets and ox-drawn carts, are demanding the right to decide how much modernization they will have, and how much they will accept from those who have made the marvelous, labor-saving, property-producing modern devices, and how much they want done in their own way, at their own pace, listening to their own hearts.

Some of their resistances will be like those of the Soviet women who successfully opposed the commands that they stop swaddling their babies with the reply, "Give us baby carriages and we will." Others will refuse the kind of school system which robs them of the help of their children (who once worked side by side with their parents in the fields) and turns their children into dropouts, unfitted for the old world or the new. But some will use the global movement for the greater education and participation of women in political decision making to reintroduce, at higher levels of planning, the human role of women in the conservation and preparation of food. Some will insist on the use of older materials close at hand, instead of foreign materials brought from a distance, and greater involvement of hand and eye in making the things they use.

The crucial difference between the historic roles of men and women was that the women waited for the life that would be created within them, while the men ranged further and further afield in search of new worlds to conquer, of new treasures to wrest from the soil, and of new secrets to tease out of the natural world. In new forms, the dilemma posed by these roles will go on. It goes on in the discussion of nuclear power and the terrible risks we are taking to get more power now, while possibly crippling generations to come. It goes on in the questions about acupuncture — surely a more "natural" method of intervention in the bodies and well-being of others than European surgery — as the new China was opened up to the curious questioning of the rest of the world.

China?

There are no photographs in this book from the People's Republic of China, although there are several pictures from Hong Kong, including the delightful one of the old women grimacing and laughing. In 1970, Ken Heyman and I asked to be allowed to visit China; I said that I would like to celebrate my seventieth birthday by a visit to China — but there was no reply. So China unvisited, reported upon by partisan well- and ill-wishers, is all that can be included in this book. When I have spoken of Moscow, or Paris, or Caracas, I have spoken of cities my eyes have seen, and cities where Ken had the opportunity to take the pictures he wanted. But for China, we have only very partial reports; neither of us has been an eyewitness in that country.

It is impossible to be certain whether the things that are being said about China today are actual reports of what has been accomplished or the expression of a prophetic sense of what some country, somewhere, must accomplish soon. For it is reported that the Chinese have found a workable compromise between nationwide planning — which went too far at first — and an older, smaller, more local scheme of things, where people live in almost self-sufficient communities, growing the food they need and making the tools they use, within a framework the outlines of which are rigorously planned and controlled.

It is said that the new China is able to feed all its people, simply but adequately, and provide them with simple clothes, which are worn by the leadership as well as by the people at large. Thus, in Bucharest, when I made a request for a visa to visit China and had tea with members of the Chinese embassy, the diplomats were dressed in their simple dark-blue working clothes. They served me tea and spoke with the extraordinary sophistication and sensitivity with which a civilization three thousand years old has provided them.

It took China a long time to come to terms with Euro-American science, technology, and social organization. Decade after decade, disorganization and famine raged through parts of this vast country, which was without the highways and railways that could have brought food from areas of plentiful harvest to regions where millions were hungry. The military strength of the West was used to establish bases for Euro-American trade and, later, by the Japanese to ravage the countryside. The people of China, like the people of Japan, went abroad to study, but unlike Japan, which has imitated Euro-American science and technology only too well and now finds itself deeply dependent upon foreign supplies of oil and raw materials, China has moved toward self-sufficiency, building new walls against too much foreign influence while using the technologies developed in the Occident selectively.

For the present, it is good for the world to cherish the belief that the Chinese, out of the wisdom of several thousand years, have, so far, found a better solution than we have, a solution in which millions of people can be an asset instead of a burden, a solution in which privilege can be used to set an example of elegant simplicity. These are the solutions toward which the older industrialized nations are groping, facing, as we do, the much older civilizations of China, India, Japan, and Southeast Asia. Our young people take tentative steps toward the contemplative life which appears and reappears in the Orient, where young people study with tremendous concentration and busy, successful men retire to a life of contrastive quiet. Some young Asians study engineering and atomic physics, but questing young Americans go to Katmandu looking for something they cannot find at home.

Whether it is a dream which only seems to be a solution, whether the simple, widely shared formulas of social participation instituted by Mao will provide a satisfying way of life for China's millions for decades to come we do not know. But the eagerness with which the rest of the world is listening stems from the belief that the Chinese have, after long travail, either developed a viable relationship between the millions who must be fed and a meaningful way of life, or, at least, are wrestling consciously with this problem.

The world is overcrowded with hungry people, and the gap between the rich and poor — between and within countries — has become unbearable, especially in a world climate of opinion which demands that all people should share prosperity equally. There is a rapidly growing sense that if all the peoples of the world are to share in what we, with our present knowledge, can produce, radical changes will have to be made. And not the least radical must be the simplification of the life style of the affluent millions in the affluent countries. Only by sharing in scarcity, with frugality, can we hope to participate together in the task of building a viable world.

Returning visitors say that the eroded hills of China have been planted with a million trees and that the whole country is turning green. In contrast, the hills of many other countries of the world are stripped of their forests, and the topsoil has been washed away. Streams are polluted from the overuse of fertilizer. Deserts are spreading wherever modern technology and the subsequent population pressures have overloaded the oases and crumbled their edges into aridity. In such a world, the vision of a million newly planted trees is heartening. It is the very magnitude of mainland China's problems and the thought that she is beginning to solve them that encourages all of us who must confront similar problems throughout the world.

12

Rethinking
the
Future

In 1945, the industrializing nations, both those who were the victors and those who were the vanquished, had more or less the same view of the world's peoples and the same panaceas for their misery. If only the whole system — literacy, medical care, automation, large-scale industry, large-scale organization, mechanization, worldwide trade, and urbanization — so very recently developed in the West, could be shared in goodwill, we would have a prosperous and happy world. When we looked at peasants and tribesmen, and at the rural poor and the urban poor, we saw them as living at the bottom of the human ladder, one which we could help them climb. With our own traditions of nationalism, of battles against foreign domination and oppression, it was easy enough to sympathize with the struggling people of the young new nations and with the oppressed people of the old nations. If every good were expressed in money, and if the per capita gross national product were the most significant measure of the state of a nation's health, then most of the peoples of the world were in dire need, but a need for which the remedy was known.

We now know that it has not turned out the way we expected; more people are suffering and enormously more are unhappy and feel aggrieved and oppressed than ever before. We are searching, blindly, for scapegoats. The big powers blame each other, the small powers blame the big powers, the economists blame the people who do not act according to economic laws, the engineers blame the people who do not seem able to learn how to use the new machines, the believers in progress blame the conservative and backward, and the believers in tradition blame the radicals, who have gone ahead too fast. Sensitive and humane critics blame the system of production, distribution, and consumption, which has been poorly designed or designed with the "wrong" set of premises in mind. Implicit in all the criticism is the assumption that if we could just deal with things on a large enough scale then we could predict and plan for all the ramifications of what we are doing. From this point of view, we might perhaps come to think of God not as white, male, and elderly, but rather as the ultimate systems theoretician and engineer.

If one looks closely, one finds that these are, in fact, all part of a general accusation of the view that the whole human problem is largely one of engineering on a planet that can be compared to a spaceship, man-made and manned by a crew who must learn to manage a complex apparatus. "Spaceship Earth" has been widely embraced as a metaphor for our human condition on this planet. The waiting eager peoples of the world are all seen as anxious to man positions on Spaceship Earth. If we used it and repaired it correctly, if we steered with care, we could journey through the cosmos in the years to come, developing greater and greater mastery. And perhaps, someday, what has already been pigeonholed as "the human component" could be completely replaced by artificial components. We might call these things "children," but they would no longer be born of men and women or subject to death. Within this engineering daydream, each of the peoples of the earth could be evaluated as more or less ready to be good crewmen; likewise, our current technology could be evaluated as more or less approaching spaceship requirements.

Edward Teller, the father of the hydrogen bomb, argued for the virtues of thorium over plutonium and how the dangers of thorium would be less. If mothers' milk was found to contain more DDT than was allowed by health department standards for cows' milk, the answer was easy enough: just give up human milk and breast-feeding, and use powdered milk. If scientific research showed that as many as 30 percent of the population of some Asian nations couldn't drink cows' milk, make a substitute. Man the toolmaker would make everything: transplant kidneys, eyes, and hearts, invent machines which would make it possible for the blind to see and the deaf to hear, reduce the problem of thinking about the future to computer simulations, sexual pleasure to a question of vibrations, birth to the supervision of test-tube technicians, birth control to monthly period extractions in which a woman need never know whether she had conceived or not. Man could do anything he wished, said Buckminster Fuller: build a dome that would cover a city or design machines so that each pupil could teach himself. Technical problems of pollution and population overload are all perfectly solvable, says Barry Commoner: all we need is a sense of social justice in peoples everywhere, along with a fair share of hardware, and their native rationality will lead them to make the right choices.

Actually, the realization that the design had failed, that women had been excluded from making it, and that the majority of the people of the world had not been consulted — these are really the same accusation, that the human qualities of human beings were ignored. And human beings were ignored because they lack the precision and reliability of machines, because considerations of love and hate and imagination make them poor systems components. Human beings are ignored when the attempt is made to eradicate the evolutionary fitness of women for bearing children and to substitute mechanical ways of making the birth of a child as independent of nature as possible. Human beings, from this point of view, are valuable only inasmuch as they have learned, from members of a very recent, newly developed kind of people — Whites — or have been educated in the schools founded to train white engineers and technologists, to manipulate the system that they have built, to make new products instead of using old ones, gradually to substitute for every natural process in the world something man-made, and, finally, to reduce each human being simply to a "human component."

When we study human history, we find how tenaciously societies cling to their solutions — the city built like a grid, the house made up of rectangular rooms, the missile that can be hurled at enemies at a distance without the danger of face-to-face combat, the armor that protects the wearer, the fortress that protects the city, the power machine that replaces the hands and skills of the worker. Seen from our modern stance, the peoples of the world who still plow with oxen, who breast-feed their children, who build their houses with their own hands — and sometimes make them round — cannot be respected. It is easy to classify together women, children, and the "lesser" breeds of people, those who are still close to the natural world through an unmediated, direct experience of the processes of growth and reproduction within their own bodies and through an overwhelming concern with the soil they till, the fields on which they graze their cattle, and the simple tools which centuries of use have shaped.

But as our knowledge has increased during this last quarter-century of frantic endeavor, it has become very clear that we have neglected essentials, that somehow in our attempt to bring an abundance of material goods to the world's people we have taken a wrong turn in our socioeconomic development, that we have made some fundamental mistakes. If we looked not only at the brief period of human civilization but at the billion years in which there has been life on this planet, at the long history of living things, we saw that over and over again, species have flourished and then declined, fallen out of step with the changing nature of the world. We found that, if we considered only the short period of civilized man and the even shorter period of our own Euro-American history with its roots in Greece and Rome, we could come to very different conclusions than if we looked at the longer history of life, just as the earth looked very different, seen from the far distances of space, than it had ever looked to the keenest geographers on the ground.

When we took a wider view, we began to look forward, to count the number of mouths that had to be fed, to measure the genetic damage that the use of nuclear power might bring, to notice that while the first generation of bottle-fed babies might fare better, the bottle-fed babies of mothers who had been bottle-fed tended to become alienated from themselves. We began to gain a new perspective of ourselves as creatures who were, after all, part of nature; we began to realize that there were many things, which had been part of our past, that we had not taken into account, and many, many things that we could hardly imagine that we should be preparing for.

The extraordinary expansion in our conception of the universe has been accompanied by an equally extraordinary shrinkage in our conception of our planet to the image of a little blue ball, isolated and vulnerable in space, alone in the solar system. This new picture of the earth is helping to concentrate our attention on those parts of the planet which have been forgotten, on the two billion people who live with many deprivations — but who have great strengths that are their legacy from the past — and on their children who have as yet no meaningful place in this changing world. If we look closely at where rebellion and the recognition of the troubles lie, we find them in the neglected parts of the world, in the tropical countries, among the peoples whom the modern world has passed by or seen only as exploitable producers of raw materials. We also find them among women who are simultaneously deprived of their maternal functions and induced to enter the marketplace to compete with men in exclusively male terms, among the young who treat the ivied halls, within which they have been forced to learn about a too narrow past, as prison walls.

The rebellion takes the same double form that rebellions have always taken — some of those who challenge the present state want nothing more than to identify with the aggressor, to become exploitative and tyrannical in turn. So, new nations want armies and fighter planes, first to defend and then to attack, just as they have been attacked; women want the right to enter male professions, which they have pronounced to be grievously wanting in humanity, and, upon succeeding, frequently become as inhumane and competitive as the men they have denounced. Young people challenge those who pollute the earth, but throw beer and soft drink cans on park grounds. Just as the leaders in the new countries have demanded the right to exploit as they were exploited, at every turn, those who challenge and question the established system seem to weaken their cause by speaking with the voices of their oppressors. By setting up a view of the world in which all are placed on a single scale, we have put each newcomer in the position of the freshman who will soon be a second-year student who can then haze and abuse and tyrannize in his turn. It is this tendency to identify with the aggressor which has made so many revolutions into tyrannies, so many movements that had their roots in a genuine rejection of injustice contribute, not to a growing humanity of a wider system, but to a growing inhumanity.

But just as the oppressed only too often identify with the oppressor, so equally others, in resonating to the cries for pity and justice and compassion, have indiscriminately identified with the oppressed. Thus we find many of the critiques which might have introduced more balance into a social system are so overstated by those whose consciences reproach them that these critiques actually reinforce the system that they are designed to criticize and change.

But there is, I think, a new way out. We have, as we never had before, ways of hearing what other people say, of watching their faces as they suffer and ask for help. The series of human inventions (begun long ago when script was invented, giving us a way to preserve the past), novel writing and theater (which brought readers and audiences portrayals of others than themselves), realistic painting, photography, film, radio, and now television, have expanded our horizons as never before. For the first time in human history, every kind of human experience can be brought within our ken: the open heart operation, the birth of a baby, the secret, sacred gestures of the priest at the altar, the agonies of the mental patient, the distorted anger of the pathological murderer, the aging bodies of people of other races, the fears and hopes of men trapped in a mine, or on a sinking ship, or in a burning building, or in danger on a journey to the moon.

Because of these new ways of recording and communicating, we have an opportunity to correct our mistakes, to take another breath, and, with a wider understanding of the needs of the whole world than human beings have ever had, to begin to re-create within creation a human way of life which echoes not a few lanes and hedgerows, a few miles of reef where seashells are gathered, nor a mountaintop where a few cling to a precarious existence, but a planet whose boundaries we have come to know, whose inhabitants we have seen, and whose lives we have begun to bring into our imaginations as never before.

But what we need most are ways of thinking and acting through which we will not be overwhelmed by increasing scale, nor driven to despair or cruel indifference by the magnitudes with which we must deal. This is a central problem for us who are conscious and sentient beings, living in the midst of an evolutionary process with a beginning on this planet which we can only guess at and a future for which we must take responsibility without being able to know in what direction it may go. The importance of each daily act — rising in the morning to the day's tasks, spending an hour explaining to a sick child in a hospital how a caterpillar turns into a butterfly, sending a few dollars to relieve famine in another part of the world — have to be brought into relationship with our decisions to build more and more missiles or to consign a million tons of grain to one country instead of another. If we can see the Nigerian fishermen with their age-old nets, the Balinese farmer plowing his land for the planting of rice, and the hungry child in India, all as parts of a whole of which we are, not distant spectators, but a part; if we can see not only power-hungry manipulators, on the one hand, and oppressed and dying people, on the other, but both — and all — caught in the same moment in history, a moment when the whole future of life on the earth, and even possibly life in the galaxy, or the universe, is at stake, the very vastness of the process can ennoble the smallest hope, deflate the most grandiose dream of world dominion, and reduce species-wide guilt to human scale, making possible action by human beings on behalf of human beings.

If we can listen to each protesting voice, not in the way it echoes our weaknesses, but in the way it points to something that is lacking in our present world view, something that should be taken into account, then no large-scale event will be too overwhelming to think about, and no newborn child too small and insignificant to protect. But this cannot be done if we use our failures, our false calculations, our misplaced confidence that technology alone would solve all our problems, to reject and decry all that human beings have been able to do since their earliest days on this earth. The impulse that has led to the modern hospital is the same impulse that a hundred thousand years ago led human beings to care for a sick member of their group, to carry a wounded person with them, or to leave food and weapons beside someone who had to be abandoned because the others were too weak to carry him. The impulse is the same, but now much more effectively implemented by the inventions that human beings have made, permitting us to transport patients in ambulances, to give them new blood, to pump oxygen into their lungs, and to treat them with powerful, lifesaving drugs.

Each enlargement of our lives that a technical discovery has brought us has also enlarged the human spirit. More human beings can be thought of as brothers whom it is murder to kill; more men and women can be thought of as human like oneself, and thus available for marriage. Building high towers, venturing into the skies, exploring space, while always carrying the danger of blasphemy, yet have made us capable of a greater humility than before. We have today a world, because its interconnectedness is only partially realized, in which larger numbers of people suffer than ever before. But it is also a world in which larger numbers of people know about that suffering and can be concerned about those who are suffering. We have today a world in which technicians attempt to imitate the bumping head of a calf against its dam or the human mother's heartbeat for infants isolated in a hospital nursery. And, it is clear, we have gone too far with this sort of artificiality in the belief that there ought to be mass-produced substitutes for every aspect of human life. But the hopes that have been generated are nonetheless hopes that were not even possible to human beings even a short hundred years ago.

If in response to the anger of the poor, of the old, of the young, of women, and of those in the disadvantaged countries our hearts are hardened because we hear in their voices only echoes of our own, we will fail. No people have ever passed through such a period as the last twenty-five years, with instant knowledge of battles lost, ships sunk, cities demolished, and thousands dying in the streets. And the next quarter-century will be as difficult a period as the last. For the future, our one sustaining strength will be a sense of our common humanity. Today, the most crucial decisions in the world are not being made by those who are most involved in the intimate and perennial concerns of human beings but by those who are closest to the hope of technological solutions. Somehow these two must be brought together, so that we can cherish the ancient knowledge of the Asian peasant, the Portuguese fisherman, and the African camel herder, and use our common humanity to translate this older wisdom from smaller and simpler scenes into a design for the whole planet, itself unified by the inventions we have made to circumnavigate it and to flash news from one side of it to the other.

Can we do it? Can we do it in time? That is both the question and the answer. If we were not conscious that the time is short to protect the people of this earth and all the life on it from disaster, there would be little possibility of setting to work hard enough. We can rebel against being identified by digits in a telephone system that works on numbers, or we can rejoice that instead of twenty telephone systems we have only one and can reach our fellows in Paris or Hong Kong quickly. The very magnitude of the task with which we are involved today may be used to reduce every human being to a cipher, or to ennoble each as a part of a whole larger than we have known before.

Background Reading

Abbott, Walter M., S.J., ed. *The Documents of Vatican II*. New York: The American Press, 1966.

Arensberg, Conrad M. "Upgrading Peasant Agriculture: Is Tradition the Snag?" *Columbia Journal of World Business*, vol. 2 no. 1 (January–February, 1967).

Bateson, Gregory. *Steps to an Ecology of Mind*. San Francisco: Chandler, 1972.

Bateson, Gregory, and Mead, Margaret. *Balinese Character: A Photographic Analysis*. 1942. New York: New York Academy of Sciences, 1962.

Bateson, M. Catherine. *Our Own Metaphor: A Personal Account of a Conference on the Effects of Conscious Purpose on Human Adaptation*. New York: Knopf, 1972.

Blum, Ralph. *Old Glory and the Realtime Freaks*. New York: Delacorte, 1972.

Boserup, Ester. *Woman's Role in Economic Development*. New York: St. Martin's Press, 1970.

Carothers, J. Edward; Mead, Margaret; McCracken, Daniel D.; and Shinn, Roger L., eds. *To Love or to Perish: The Technological Crisis and the Churches*. New York: Friendship Press, 1972.

Carson, Rachel L. *Silent Spring*. Boston: Houghton Mifflin, 1962.

Commoner, Barry. *The Closing Circle*. New York: Knopf, 1971.

Doxiadis, C. A., ed. *Anthropopolis: City for Human Development*. Athens: Athens Publishing Center, 1974.

Doxiadis, C. A., and Papatoannou, J. G. *Ecumenopolis: The Inevitable City of the Future*. Athens: Athens Center for Ekistics, 1974.

Dubos, René. *Beast or Angel?* New York: Scribners, 1974.

Eisenberg, J. F., and Dillon, Wilton S., eds. *Man and Beast: Comparative Social Behavior*, pp. 369–81. Washington, D.C.: Smithsonian Institution Press, 1971.

Ekistics, published by Athens Center for Ekistics of the Athens Technological Organization, 24 Strat, Syndesmou Street, Athens 136, Greece.

Environment, published by the Scientists' Institute for Public Information, 438 N. Skinner Blvd., St. Louis, Missouri, 63130.

Erikson, Erik H. *Gandhi's Truth: On the Origins of Militant Nonviolence*. New York: W. W. Norton, 1969.

Frank, Lawrence K. *Nature and Human Nature: Man's New Image of Himself*. New Brunswick: Rutgers University Press, 1951.

Golding, William. *Inheritors*. New York: Harcourt Brace Jovanovich, 1962.

Gorer, Geoffrey. *The Danger of Equality*. London: Cressett Press, 1966.

Gottmann, Jean. *Megalopolis*, 1961. Millwood, N.Y.: Krauss, 1974.

Hall, Edward T. *The Hidden Dimension*. Garden City, New York: Doubleday, 1966.

Hersey, John. *Hiroshima*. New York: Knopf, 1946.

Land, George T. *Grow or Die: A Unifying Principle of Transportation.* New York: Random House, 1973.

Mailer, Norman. *Of a Fire on the Moon.* Boston: Little, Brown, 1971.

Mead, Margaret. *Culture and Commitment: A Study of the Generation Gap.* Garden City, New York: Natural History Press & Doubleday, 1970.

————. *New Lives for Old: Cultural Transformation — Manus, 1928–1953.* New York: Morrow, 1956.

Mead, Margaret, ed. *Cultural Patterns and Technical Change.* Paris: UNESCO, 1953.

Mead, Margaret, and Heyman, Ken. *Family.* New York: Macmillan, 1965.

Meadows, Donella; Meadows, Dennis L.; Randers, Jørgen; and Behrens, William W., III. *The Limits to Growth.* Washington: Potomac Associates, 1972.

Mesarovich, Mihajlo, and Pestel, Eduard. *Mankind at the Turning Point.* New York: Dutton, 1974.

Mosley, J. Brooke. *Christians in the Technical and Social Revolutions of Our Time.* Cincinnati: Forward Movement Publications, 1966.

Niehoff, Arthur H., ed. *A Casebook of Social Change.* Chicago: Aldine Press, 1966.

Oltmans, Willem L., ed. *On Growth: The Crisis of Exploding Population and Resource Depletion,* pp. 18–25. New York: Capricorn Books/Putnam, 1974.

Perloff, Harvey S., and Sandberg, Neil C., eds. *New Towns: Why and for Whom?* pp. 117–29. New York: Praeger, 1973.

Poggie, John J., Jr., and Lynch, Robert N., eds. *Rethinking Modernization: Anthropological Perspectives,* pp. 21–36. Westport and London: Greenwood Press, 1974.

Saturday Review/World. New York, N.Y.

Tillion, Germaine. *Algeria: The Realities.* New York: Knopf, 1958.

Tiselius, Arne, and Nilsson, Sam, eds. *The Place of Value in a World of Facts,* pp. 419–27. New York: Wiley Interscience Division, 1970.

Turnbull, Colin. *The Lonely African.* New York: Simon & Schuster, 1962. Reprinted, Garden City, New York: Doubleday, 1963.

United Nations, Conference on the Human Environment, *Report,* Stockholm, June 5–16, 1972. (A/CONF. 48/14; Sales No. 73.II.A.14.), 1972.

United Nations, Secretary-General. World Population Conference. Note. October 2, 1974 (E/5585.).

Waddington, C. H., ed. *Biology and the History of the Future.* Edinburgh: Edinburgh University Press; Chicago: Aldine-Atherton, 1972.

————. *The Ethical Animal.* London: Allen and Unwin, 1960; New York: Atheneum, 1961.

Ward, Barbara. *Human Settlements: Crisis and Opportunity.* Prepared for Habitat–United Nations Conference on Human Settlements, in press.

Ward, Barbara, and Dubos, René. *Only One Earth: The Care and Maintenance of a Small Planet.* New York: W. W. Norton, 1972.

Weiner, Norbert. *Human Use of Human Beings.* Boston: Houghton Mifflin, 1954.

White, Lynn, Jr. *Medieval Technology and Social Change.* Oxford: Oxford University Press, 1966.

Wolfenstein, Martha. *Disaster.* Chicago: Free Press, 1957.

World Council of Churches. *Christians in the Technical and Social Revolutions of Our Time,* World Conference on Church and Society. Geneva, July 12–26, 1966, The Official Report. Geneva: World Council of Churches, 1967.

Worth, Sol, and Adair, John. *Through Navajo Eyes.* Bloomington: Indiana University Press, 1972.